# Colonial Pathologies, Environment, and Western Medicine in Saint-Louis-du-Senegal, 1867–1920

# Society and Politics in Africa

Yakubu Saaka
*Founding Editor*

Akwasi Osei
*General Editor*

Vol. 21

PETER LANG
New York • Washington, D.C./Baltimore • Bern
Frankfurt • Berlin • Brussels • Vienna • Oxford

Kalala Ngalamulume

# Colonial Pathologies, Environment, and Western Medicine in Saint-Louis-du-Senegal, 1867–1920

PETER LANG
New York • Washington, D.C./Baltimore • Bern
Frankfurt • Berlin • Brussels • Vienna • Oxford

Library of Congress Cataloging-in-Publication Data
Ngalamulume, Kalala J.
Colonial pathologies, environment, and Western medicine
in Saint-Louis-du-Senegal, 1867–1920 / Kalala Ngalamulume.
p. cm. — (Society and politics in Africa; v. 21)
Includes bibliographical references and index.
1. Tropical medicine—Senegal—History. 2. Tropical medicine—Social aspects—
Senegal—History. 3. Public health—Senegal—History—19th century.
4. Epidemics—Social aspects—Senegal—History.
5. France—Colonies—Africa, French-speaking West—Social conditions.
6. French—Senegal—History—19th century.
I. Title. II. Series: Society and politics in Africa; v. 21.
RC962.S36N45   616.9'883009663—dc23   2012009354
ISBN 978-1-4331-1499-1 (hardcover)
ISBN 978-1-4539-0813-6 (e-book)
ISSN 1083-3323

Bibliographic information published by **Die Deutsche Nationalbibliothek**.
**Die Deutsche Nationalbibliothek** lists this publication in the "Deutsche
Nationalbibliografie"; detailed bibliographic data is available
on the Internet at http://dnb.d-nb.de/.

© 2012 Peter Lang Publishing, Inc., New York
29 Broadway, 18th floor, New York, NY 10006
www.peterlang.com

All rights reserved.
Reprint or reproduction, even partially, in all forms such as microfilm,
xerography, microfiche, microcard, and offset strictly prohibited.

To Yaya Vicki Bibole, Tatu Kande, Muty
Lina, Herve, Carole, and Faziah;
and in memory of my father, mother, and uncle
for all the sacrifice

# Contents

|  |  |  |
|---|---|---|
|  | List of Illustrations | ix |
|  | Acknowledgments | xi |
|  | Abbreviations | xv |
|  | Introduction: Setting, Themes, and Conceptual Perspectives | 1 |
| 1. | The Making of a Colonial City | 16 |
| 2. | The "White Man's Disease": The Great Yellow Fever Epidemics, 1867–1900 | 50 |
| 3. | The "Black Man's Disease": Cholera and Social Inequality, 1868–1899 | 88 |
| 4. | A Conflict of Interests among Commerce, Competing Conceptions of Public Health, and Civil Liberties, 1882–1901 | 120 |
| 5. | The Scientific Missions to Senegal and Brazil and the New Paradigm, 1901–1912 | 164 |
| 6. | Plague and Violence in Saint-Louis-du-Senegal, 1917–1920 | 185 |
|  | Conclusion | 208 |
|  | Sources | 211 |
|  | Index | 231 |

# Illustrations

## MAPS

| | | |
|---|---|---|
| 1. | Senegal | 6 |
| 2. | Saint-Louis | 26 |
| 3. | St. Louis and Isolation Camps | 75 |
| 4. | The Evolution of the Bubonic Plague Epidemic in Saint-Louis | 198 |

## FIGURES

| | | |
|---|---|---|
| 1.1 | Saint-Louis Population Growth, 1787–1921 | 20 |
| 3.1 | Daily Reported Cholera Deaths, Saint-Louis, 1868 | 99 |
| 3.2 | Daily Reported Cholera Deaths, Saint-Louis, 1893 | 113 |
| 6.1 | Evacuations of the Residents of Guet-Ndar, Saint-Louis, 1918 | 201 |

## TABLES

| | | |
|---|---|---|
| 1.1 | Ethnic/Racial Distribution of the Population (no Troops or Navy) | 21 |
| 1.2 | Epidemics and Diverse Scourges, Senegal, 1859–1927 | 34 |
| 1.3 | The Medical Staff in Senegal, 1857–1867 | 38 |

| 1.4 | The Medical Staff in Senegal, 1873–1894 | 39 |
| 1.5 | Distribution of Catholic Nuns in Medical Facilities in Saint-Louis | 40 |
| 2.1 | Yellow Fever Mortality in Saint-Louis in 1881 | 62 |
| 2.2 | Recapitulation: Official Mortality from Yellow Fever and Cholera, 1867–1881 | 63 |
| 2.3 | The Quarantines and *Cordons Sanitaires* in Saint-Louis | 76 |
| 3.1 | Changing Level of Salt in the Senegal River, Saint-Louis, 1851–1852 (per 100 gr.) | 94 |
| 3.2 | Distribution of Water to Government Employees, Saint-Louis, 1873 | 95 |
| 3.3 | Daily Reported Cholera Deaths, Saint-Louis, 1868 | 97 |
| 3.4 | Distribution of Deaths from Cholera, Saint-Louis, 1868 | 98 |
| 3.5 | The Revised Estimate of the Badois Contract (in francs) | 107 |
| 3.6 | The Cost of Bathing at the Military Hospital, Saint-Louis, 1892 (in francs) | 109 |
| 3.7 | Daily Reported Cholera Deaths, Saint-Louis, 1893 | 112 |

## Plates

| 1. | The Faidherbe Bridge next to the old bridge | 154 |
| 2 | A street in Saint-Louis | 155 |
| 3 | A street in Saint-Louis | 156 |
| 4 | A slum area at the south end of the city island | 157 |
| 5 | A view of Guet-Ndar slum | 158 |
| 6 | The small arm of the Senegal River | 159 |
| 7 | A mule-drawn cart in Ndar-Toute quarter | 160 |
| 8 | The Grotte de Lourdes in Sor | 161 |
| 9 | The monument to the victims of the 1878 yellow fever epidemic | 162 |
| 10 | Water tank in Sor | 163 |

# Acknowledgments

This book would not have been possible without the assistance of many people and institutions. The initial fourteen months of major fieldwork and archival research in Senegal and France was assisted by grants from the Rockefeller Foundation and from the joint Committee on African Studies of the Social Science Research Council and the American Council of Learned Societies with funds provided by the Ford, Mellon, and Rockefeller Foundations. I would like to take this opportunity to thank them for their generous support. Subsequent research trips to Aix-en-Provence and Paris were made possible thanks to the grants from the Faculty Grants and Awards, the Africana Studies program, and the Center for International Studies at Bryn Mawr College. I am very grateful to Provost Kimberly Wright Cassidy for covering the author's contribution to the cost of publication of the book and for her generosity. I wish to thank Bryn Mawr College President Jane McAuliffe and former President Nancy Vickers for all of their support. My gratitude is also extended to the faculty and staff of the Africana Studies Program and of the History Department at Bryn Mawr College for their constant support and friendship. I also wish to thank the staff in the Provost's Office, especially Nona Smith and Elizabeth Shepard-Rabadam for their help and friendship. I would like to thank Dianna Xu and her former students Corey Norcross, Andrea Smilie, and Jaclyn Lang in the Computer Science Department as well as Del Ramers from the Visual Resources for their help with the graphs and maps.

A great debt of gratitude is owed to those who helped me during my fieldwork and archival research. In Senegal, I am indebted to Charles Becker and his family for their support during my stay in Dakar. Both Charles and René Collignon provided clues to key sources in the history of health and medicine in Senegal. I also thank Professors Penda Mbow, Babacar Fall, Mariama Mbow, Tahiru, Mohamed Mbodj, and Malik Ndiaye for their hospitality and for providing useful contact information. I would like to thank Dr. Saliou Mbaye, Mamadou Ndiaye, Ngor Sene, and all the staff of the National Archives of Senegal as well for their help. I owe thanks to Angelique Diop for sharing her perspectives on health issues in Senegal, to Pauline Kemayi Nkitabungi and Evelyne Baumann for their support and friendship. In Saint-Louis, I wish to thank the faculty and staff at the Université Gaston Berger. I owe a great debt of gratitude to former Rector Ahmadou Lamine Ndiaye and Professors Babacar Kante, Maweja Mbaya, and Muamba Cabakulu for their assistance and friendship while doing research in Saint-Louis.

In France, I wish to thank Professor Jean-Louis Triaud at the Université de Provence for his help with visa and accommodation, and for his friendship. My thanks to Jean-Paul Bado and his family for their hospitality and for the insightful suggestions Jean-Paul provided. I am very thankful to the staffs of the following archives and libraries for their assistance: the Centre des Archives d'Outre-Mer in Aix-en-Provence, the Académie de Médecine in Paris, the Service Historique de la Marine in Vincennes, the Bibliothèque Nationale de France in Paris, the Archives de la Congrégation du Saint-Esprit in Chevilly-Larue, the Archives de la Congrégation Saint-Joseph de Cluny in Paris, and the Institut Pasteur in Paris.

Some of the material on yellow fever (chapter 2) were published in "Keeping the City Totally Clean: Yellow Fever and the Politics of Prevention in Colonial Saint-Louis-du-Senegal, 1850–1920,'" *Journal of African History*, 45.2 (2004). Portions of chapter 3 previously appeared in "Coping with Disease in French Empire: The Waterworks in Saint-Louis-du-Senegal," in *African Water Histories: Transdisciplinary Discourses* (Vanderbijlpark, South Africa: Vaal Triangle Faculty, North-West University, 2005). I have previously published chapter 6 in *Cahiers d'Études Africaines* XLVI (3), 183, 2006. I wish to thank the publishers of these materials for their permission to reprint them.

At various stages of the preparation of this research project, many colleagues have read and made critical observations along the way. I am deeply grateful to my mentor and friend, David Robinson, for his continuous support and for being an inspiring role model of intellectual rigor and compassion. I also gratefully acknowledge the comments made by Lee Casanelli, Martin Klein, Myron Echenberg, Madhavi Kale, Sharon Ullman, and Bob Washington. I also wish to thank the anonymous readers of the *Journal of African History* and the Cahiers d'Études Africaines, and reviewers and to Phillis Korper, editor of the Peter Lang, for her suggestions. Many thanks to Jackie Pavlovic and her staff at Peter Lang for their formatting and bookmaking skills. I also would like to thank my 'Senegalisant' colleagues and friends for their stimulating conversations: Ghislaine Lydon, Hilary Jones, Cheikh Babou, and Ellen Foley.

In Congo-Kinshasa, I owe a special debt of gratitude to Muanangana (King) Emery Wafuana Kalamba for all of his material and moral support and his friendship, and to Sylvain Kayembe also in Kinshasa for his assistance. I wish to thank Professor Munayi Muntu Monji for providing guidance early in my career, and Professors Tshisungu Lubambu, Kambayi Buatshia, Ndaywel E Ziem, Sabakinu Kivilu, and Mutamba Makombo and Chef des Travaux Motena Mokaba in Kinshasa for their intellectual engagement and friendship.

Finally, I should like to thank my family in the United States, Africa, and Europe for their unwavering love, support, and for the sacrifice they have made.

# Abbreviations

| | |
|---|---|
| ACSE | Archives of the Congrégation du Saint-Esprit (Chevilly-Larue, France) |
| AIP | Archives de l'Institut Pasteur (Dakar, Sénégal) |
| AMI | Assistance Médicale Indigène |
| ANF | Archives Nationales de France (Paris) |
| Ann d'Hygy. Col. | Annales d'Hygiène Coloniale |
| ANS | Archives Nationales du Sénégal (Dakar) |
| ANSOM | Archives Nationales, Section Outre-Mer (Aix-en-Provence) |
| AOF | Afrique Occidentale Française |
| AOM | Archives d'Outre-Mer (Aix-en-Provence) |
| BA | Bulletin Administratif |
| BO | Bulletin Officiel |
| CAOM | Centre des Archives d'Outre-Mer (Aix-en-Provence) |
| CHPS | Council of Hygiene and Public Salubrity |
| FWA | Federation of French West Africa |
| ISC | International Sanitary Conference |

# INTRODUCTION

# Setting, Themes, and Conceptual Perspectives

I arrived in Dakar, capital of Senegal, on New Year's Eve 1993 to start a fifteen-month research that would later take me to France and back to Dakar and Saint-Louis. By early June, the majority of the French expatriates I had met were making preparations to leave for France, just before the arrival of the *hivernage*, the hot and humid season, which in their minds was still associated with heat and disease. Eye-witness accounts from the 1970s have also noticed the same response. It has been reported that French expatriates in Senegal received not only twice the salaries their counterparts with the same rank were given back home in France but also hardship bonuses for working in what the central administration in Paris perceived to be a dangerous disease environment. So the persistence of the negative images of Senegal and the continuity of political and bureaucratic responses to them was puzzling and could be traced back at least to the second half of the nineteenth century.

*Colonial Pathologies, Environment, and the Western Medicine in Saint-Louis-du-Senegal, 1867–1920* analyzes Saint-Louis' experiences with sanitation and epidemics, especially yellow fever, cholera, and bubonic plague epidemics in the second half of the nineteenth century and in early twentieth. Although the study is about Saint-Louis, it is also about Senegal—the two meant the same thing for most of the nineteenth century—and the ways in which the colony was conceptualized and contextualized by French travelers, medical doctors, and colonial officials during its formative years as a land of heat, filth and foul odors, disease and death that was unsuitable to French settlement, and how the perceptions they had of the land and the native populations affected the government's decisions, not just health policies. So this book is about the challenges the French authorities faced to make Senegal a healthy colony and Saint-Louis the healthiest port city/capital and about the politics of public health. It provides a valuable contribution to our understanding of the urban processes and the daily life in a colonial city during the formative years of the French empire in West Africa. In doing so the book also addresses broader issues that other historians have paid attention to in the past two decades, such as the social construction of disease, the relationship between disease and imperialism, the 'colonization of the body,' and the responses to health policy.[1] This study explores the variations and differences of these processes in the context of colonial Saint-Louis-du-Senegal.

## Themes

Five dominant themes run across chapters of this book. The main theme deals with the ways in which the French public health authorities constructed medical knowledge about the etiology and epidemiology of diseases, the landscape, and the Africans, how they clung to this older explanatory model despite the challenge brought about by the supporters of bacteriology from the mid-1880s on, and how the paradigm shift took place. A related theme focuses on the link between the emergence of Senegal's 'hot climate' cliché and the increasing role of the physicians within the colonial administration. A third theme examines the stigmatization of the urban poor. A fourth theme relates to the responses of various segments of the urban population to the sanitary and medical measures during the epidemic crises. The last theme deals with the urban residential segregation policies that were recommended as the permanent solution to the persistent 'native problem' associated with the re-emergence of epidemic diseases.

One of the main arguments in this book is that the outbreaks of major epidemic diseases gave way to the epidemic of fear and suspicion, to the epidemic of stigmatization of the *indigènes*, and to the adoption of drastic and disruptive sanitary measures that led to a conflict between the interests of competing conceptions of public health and those of commerce, civil liberties, and popular culture. Indeed, the French authorities had generated a great deal of 'knowledge' about the landscape on which Saint-Louis was built and about the Africans to which infectious diseases were ascribed. Such knowledge making was based on the existing medical theories and shifting etiologies and treatments, independent observations in the colony as well as the images and ideas about the 'exotic' environment, and essentialist assumptions, and it resulted in the myth of insalubrious African climate and natural environment as one of the leading causes of ill health that occupied French imagination most of the nineteenth century.

By the 1880s the influence of medical doctors within the colonial administration had greatly increased and their 'scientific' opinions were sought, especially concerning medical and sanitary measures, settlement schemes, and the length of service and regular furloughs of civil servants and European troops that were central to the survival of Europeans in a perceived threatening environment. Physicians owed this ascendency in part to the epidemic of fear that 'colonial' pathologies and the hot climate provoked. The list of negative effects of the hot climate on the bodies of the colonizers terrified anyone who contemplated a future in the colonies. The 'tropical anemia' resulting from even a short stay in a 'warm climate' was the most feared condition. E. Aubert, one of the contributors to the *Annales coloniales*, reported in 1901 that

> This (intertropical) climate exercises, indeed, an anemic, debilitating action on the European who exposes himself to it. All the physicians who have sojourned in the colonies have observed that anemia. Paleness of the skin, painful digestions, and vertigo associated with the defective functioning of the stomach and of the intestine, palpitations, breathlessness, (and) memory loss are the symptoms of the tropical anemia (that are) capable of appearing already within a few days or a few weeks, even before the weakened organism presents any lesion whatsoever.[2]

The 'tropical anemia' became a widely used cliché in the medical and colonial discourse even if those with some colonial experience knew that these symptoms could well be associated with such conditions as hepatitis, dysentery, malaria, relapsing fever, typhoid fever, and cholera.[3] Thus, the negative image of Senegal that resulted from this kind of 'knowledge' provided the rationale for the implementation of the policies of racial residential segregation and repatriation of the Europeans during the epidemic crises as well as during the summer months or in-between epidemics. Fear of epidemic diseases was also instrumental in the determination of the duration of service for colonial officials. The perceived debilitating nature of the 'warm climates' on the bodies of the colonizers led to the development of a new business of colonial SPAS and hydrotherapy.[4] The advent of the germ theory of disease did not necessarily lead to the adoption of a more scientific approach to public health, as the chapter on the bubonic plague epidemic will show. Instead, the landscape of Senegal and the urban poor were blamed for the spread of plague. The repatriation or evacuation of the Europeans when the epidemic diseases were menacing became the archetype model of French (and Western) responses to the political turmoil or humanitarian crises in post-independent Africa in general, not just health threats.

Other environmental factors had to do with the problems that accompanied urbanization. Indeed, overcrowding, unsanitary housing shared with domestic animals, rubbish, and the lack of sewers figured in a good order among the predisposing factors that the authorities believed were responsible for the spread of diseases in the city. These factors posed a serious challenge to the colonial social order because they were associated with the presence in the urban environment of those classified as *indigènes*, some of whom were French nationals. Regular fires that destroyed their shacks provided the authorities with a good pretext to adopt a policy of forced removal aimed at eliminating them from the city center. All these factors contributed to shaping the medical discourse and practices and their changing nature over time.

Filth and dirt or uncleanness, as a by-product of the growth of Saint-Louis, came to symbolize all that was wrong with the colonial city. Anthropologist Mary Douglas has shown that particular concern over dirt—as matter out of place—arises when it becomes a threat to good order and to vulnerable relationships in a society. She underlined that 'some pollutions are used as analogies for expressing a general view of the social order,' and 'in a dialogue of claims and counter-claims to status.'[5] This statement holds true for Saint-Louis where the productive activities that made life in the city possible for the majority of city residents, such as the fishing industry, food supplies, animal uses for transportation, and animal slaughtering, created dirt and filth that threatened the urban order. The colonial authorities felt the need to protect the city from defilement. The French medical discourse constructed the Africans as filthy-prone residents, and depicted the colonial city as an environment characterized by a dangerous disorder of things out of place that poisoned it.

The association of the *indigènes* with the filthy state of the city and their stigmatization reveal the mental topography of race relations. Guet-Ndar, inhabited mostly by the fishermen of Wolof origin, became the site of dirt and filth par excellence and was thus stigmatized. In the correspondence of the colonial and medical officials, this most popular slum was depicted as a place of dis-order and a 'space of death,' to borrow Michael Taussig's expression.[6] In other words, for the French authorities, there was a cognitive dissonance in that the reality of everyday life in Saint-Louis was in contradiction with the discourses about and representations of '*l'oeuvre civilisatrice*' (or 'civilizing mission'), 'of the ways in which a "civilized" society should look and smell and its citizens act.'[7] The 'enigma of otherness' of the *indigènes* remained a recurrent theme in the medical discourse during the period under investigation. And the female *indigènes* became the ultimate 'Others' because of the attraction they supposedly exercised on the male residents and outsiders and of their status as 'special carriers' of the 'germs' of yellow fever and of sexually transmitted diseases that resulted in 'a great number' of new infections, deaths, and hundreds of days of morbidity among the soldiers, and unanticipated expenses for the administration.

Concerning the theme of urban planning in Saint-Louis, it should be emphasized that the perception of yellow fever as the 'white man's disease' perhaps more than any other factor, provided the rationale for the change from early forms of interweave or juxtaposition of administrative, commercial, and residential structures to urban residential segregation by race and class, a process that took decades to complete. The sources on yellow fever are more abundant than those dealing with any other epidemic or endemic diseases affecting the urban population, thanks to the numerous health boards that were formed in order to find solutions to disease threats. Yellow fever and the subsequent stigmatization of the urban poor not only provided compelling reasons for the colonial state massive intervention in more and more aspects of its citizens' and subjects' lives, but they also raised serious questions about the growing gap between the proclaimed objective of assimilation and everyday practices, about citizenship and civil liberties. The scope and reach of the medical and sanitary measures revealed a division among physicians, between colonial state officials and municipal authorities, and between the business community and the colonial administration as well as within the business community. The outbreak of bubonic plague in 1917 led to the evacuation of the entire population of the slum of Guet-Ndar, thus bringing the process of racial residential segregation to its climax.

Another line of argument is that, as a product of its particular history, Saint-Louis' dynamic civil society, some segments of which controlled the municipal government and the General Council and defended local interests, posed a serious challenge to racialized health policies affecting the urban poor, and to sanitary cordons and quarantine measures that threatened their economic interests.

## Conceptual Framework

The study of Saint-Louis brings together the categories and conceptual frameworks from urban history, political history, and social history of medicine.

### Urban and Political Histories

Although the unit of analysis is the city, the process of its growth cannot be understood or explained if one does not consider the city in the context of the region and the network of cities, and international relations. These perspectives can help identify the forces involved in the process of city growth and decline, such as technology, demography, and market forces; it can also help explain different functions assumed by the city. The region can be defined either by ecological factors or by economic and social organization.

Paul M. Hohenberg and Lynn H. Lees have provided a framework for analyzing the link between wider economic, political, social, and cultural processes and the microcosm of the cities. They proposed two theoretical models for approaching the study of the city: the Central Place System and the Network System.[8] The Central Place System, first formulated by the German geographer W. Christaller in the 1930s and improved by other scholars,[9] focuses on the role that the city plays in relation to its region, that is 'a central place, supplying its surrounding with special services—economic, administrative, or cultural—that call for concentration at a point in space.'[10]

This perspective helps approach Saint-Louis as a central place for other central places (marketplaces) in the Senegalo-Mauritanian zone,[11] with administrative functions slowly grafting into the economic functions. Saint-Louis' region initially included the marketplaces (*escales*) along the Senegal River, such as Fort Saint-Joseph in the Galam (1700–1732), Fort St. Pierre (1714), Matacong (1730), Bakel (1818), Dagana (1819), Richard-Toll (1820), Merinaghen (1842), and Lampsar (1843) before further expansion. During the second half of the nineteenth century the commercial hinterland was extended with the creation of new marketplaces: Podor (1854), Medine (1855), Matam (1857), Ndangan (1858), Sadlé (1859), Kenieba (1858, 1861), Aere (1866), and Kayes (1879); as well as the establishment of trade relations with Moors. Thus, all the central places in the Senegalo-Mauritanian zone formed Saint-Louis' hinterland.

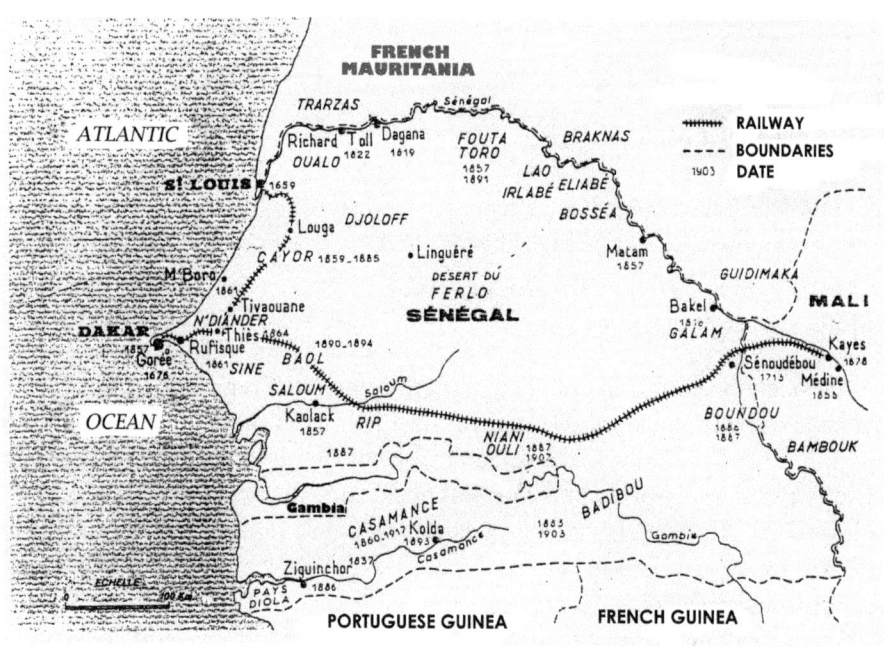

Map 1. Senegal (adapted from Niane and Suret-Canale, 1961, 109)

The second model emphasizes the link between the Senegalo-Mauritanian zone and the external world. Here the city is analyzed in the context of networks of trade, information, and influence. The model explained the importance of treating cities as part of a larger world system, a concept popularized by Immanuel Wallerstein.[12] de Vries developed the theoretical concept of 'urban system,' which includes cities that are linked by economic interactions[13] and allows us to examine the growth of Saint-Louis in relationship with other cities in the French empire and in West Africa. One of the analytical constructs in the urban system is the concept of 'colonial city' that was theorized by G. J. Telkamp in his pioneering work where he underscored the importance of situating colonial cities in in the context of broad historical processes (colonialism, capitalism, and industrialization).[14] Both Ross and Telkamp argued in favor of the study of a city's life as a way of gaining insight into the history of a colony. They proposed an analysis of the city at three levels: (1) the functions of the city from a political economy approach that emphasized the complex inter-penetration of the city, the countryside, and the world economy; (2) the organization of the city, or the way it was controlled; and (3) the spatial layout of the city, including its social stratification.[15] Ever since, there has been a growing interest in the study of colonial cities, including colonial port cities, as dynamic forces of social change in Africa.[16]

These approaches to the study of cities, in general, and colonial cities in particular provided insights into the investigation of Saint-Louis as a colonial port-city/capital. For, as a port-city, Saint-Louis served the function of gateway for the export of colonial wares and the import of goods from Europe. It functioned as the headquarters of firms involved in trade and the services associated with it. Like in all colonial cities, the essential of the economic, social, and political powers were concentrated in the hands of the French elite, who were racially, culturally, and religiously different from the indigenous population. Also Saint-Louis' tertiary sector was important because of the focus on trade. The Creoles played a major role in the political and economic institutions of the colony during the formative years of the French West Africa.

Thus, the growth of Saint-Louis cannot be comprehended if one focuses on the city alone or simply on its role as the capital of Senegal because, for decades, Saint-Louis was the privileged link in international networks, including Europe, Africa, and America. Its centrality explains the number of entities of which it was the acknowledged capital: French Senegal (1848–1960), French West Africa (1895–1904), and Mauritania (1904–1957). To many travelers in francophone West Africa the trip to Saint-Louis was an inland voyage as opposed to one abroad. Many within the francophone West African elite were trained in Saint-Louis. The Network System includes cities which are cosmopolitan, pluralistic, and heterogenetic, diffusing the dominant culture to their surroundings. Saint-Louis had all these characteristics.

Scholarship on Saint-Louis as a colonial city is uneven. Most studies have examined the origins and growth of Saint-Louis, and the political, administrative, military, and cultural functions that the city fulfilled in the nineteenth and twentieth centuries,[17] the urban elite,[18] urban institutions,[19] and urban architecture.[20] These studies help understand aspects of Saint-Louis' life but they do not focus on the impact of epidemic diseases on these urban processes.

## Social History of Medicine

Since K. David Patterson's observation in 1974 about African medical history as an untouched field, an increasing number of scholars have been focusing their attention on the relationship among disease, Western medicine, and colonial rule in Africa.[21] Thematically, the main concerns have related to the role of colonial medicine in the conquest of Africa, the social history of several 'colonial diseases,' disease and urban segregation, disease ecology and political economy, medical ideologies, and medical profession. Interpretations have also varied. Although most historians have looked at biomedicine as a tool of empire and a means of social control, they have disagreed on its effectiveness. Conservative interpretations have presented the history of Western medicine in Africa as an account of its successes in reducing mortality and in achievement population growth. Liberal interpretations have tried to show the record of colonialism on health issues was a balanced one. Radical interpretations have emphasized racialized health policies, health inequalities, and the introduction of new diseases.

However, these studies do not focus on disease, health care, and medicine in Saint-Louis. Only one scholarly article by Claude Pulvenis (1968) discussed the impact on the city of the 1881 yellow fever epidemic, which struck both Saint-Louis and the troops involved in the conquest of the Upper Sudan. But his study is limited in scope to one epidemic crisis and does not provide a global picture of the incidence of the major epidemics in the city.[22] In a bibliography on health and population in Senegambia published two decades ago by René Collignon and Charles Becker (1989), one can find studies related to Saint-Louis written by colonial officials, physicians, and missionaries. But they are primarily descriptive and emphasize epidemiology and health efforts of the colonial administration. They do not study the epidemics in relation to the political, social, and economic realities of the time. Even Philip Curtin, who has made an important contribution to the history of Western medicine in the 'tropical world,' did not include Saint-Louis in his seminal article on disease, medical knowledge, and urban planning in tropical Africa. A well-known historian of public health of Senegal, Myron Echenberg recently pointed out that it is strange that 'given Senegal's prominence in Francophone African historiography generally, no full-length studies of the history of health and disease in this country have yet been attempted.'[23] The scarcity of case studies accounts for this lacuna. Echenberg's well-researched book focuses on plague in Dakar and its hinterland between 1915 and 1945. Yet, Saint-Louis was the first experimental terrain of French colonial policies that were later implemented in French West and Equatorial Africa. My study dealing with yellow fever, cholera, and bubonic plague in Saint-Louis will fill the gap and be a good addition.

The approach to *Colonial Pathologies, Environment, and Western Medicine* has also been inspired by David Robinson's perspectives on hegemony and resistance[24]; Charles Rosenberg's influential work on the social and political construction of epidemics[25]; Michel Foucault's analysis of 'biopolitics' or 'biopower'—meaning a technology of power separate from disciplinary mechanisms or tools—and discoursive practices[26]; Alain Corbin's work on the smell of city's streets[27]; the French *Annales*' insights on the subjective perceptions and capacities of action of social actors (agency) and on their use of Italian-inspired micro-history techniques[28]; Michel de Certeau's insights on the disruptive presence of the 'other' in systems of power and thought[29]; and works on symbolic anthropology.

## Sources

The official archival documentation for this book is drawn from the National Archives of Senegal (A.N.S.) in Dakar, the Municipal Archives in Saint-Louis, the National Archives of France, Section d'Outre-Mer (A.N.F.O.M.), in Aix-en-Provence, and the Navy Historical Service in Chateaux de Vincennes. The official sources contain a great deal of data on the epidemics as well as the routine health care. But the statistics concerning the diseases treated in the hospitals and dispensaries are problematic because many disease categories are vague and ambiguous until the advent of bacteriology. The data provide more details about the Europeans and *métis* than about the *indigènes*. Before the influx of more Europeans into Saint-Louis in the 1890s, most medical records are silent about women (wives, mothers, daughters, and relatives, and prisoners), with the exception of *filles soumises* ("submissive girls"), a euphemism for prostitutes who were in contact with soldiers, and an impact on their health. Mortality statistics are inadequate as they record only the deaths that occur in the hospitals.

The official documents provide the researcher with interesting insights into the health care delivery system and various aspects of urban life. The documents contained, nevertheless, rather moral judgments about those classified as *indigènes*. Indeed, the majority of health officers viewed Western medicine as an integral part of the "Civilizing Mission," which sustained the official ideology, and hospitals as "healing machines" ("machines à guérir"), to borrow Michel Foucault's expression.

Physicians perceived themselves as agents of "progress" engaged in a struggle against the forces of darkness and barbarity and in a mission to "eradicate" diseases. Few surgeons and physicians really tried to understand the local culture, lifestyle, and medical systems not only for ideological reasons but also for other reasons that have to do with their training and time constraints. Of the fifteen courses that were taught in the Navy surgery schools at Brest, Rochefort, and Toulon in the 1870s, not one was devoted to the study of the diseases that prevailed in Africa or Asia or to indigenous medical systems. While the expression "exotic pathology" appears twice in the course description, it is only one chapter in a course dealing with other issues.[30] The creation in April 1890 of the *Ecole du Service de Santé de la Marine*, affiliated with the *Faculté de Bordeaux*, was motivated in part by the felt need to remedy the situation.

Moreover, surgeons and physicians did not stay long enough in their positions to gain the necessary experience and increase their knowledge about local realities. They were appointed for two or three years, and they left before becoming familiarized with the local environment. It was only after the creation of the *Assistance Médicale Indigène*, in 1905, that a few medical doctors extended their stay by transferring into the new bureaucracy.

The official archives also contain the letters and petitions written by the Senegalese that reveal the reactions and initiatives of some groups or individuals to different sanitary and medical measures imposed upon them by the colonial government.

I have also consulted the Archives of the Congregation of Saint-Esprit (ACSE) in Chevilly-Larue (France) and of the Sisters of Saint-Joseph de Cluny housed in Paris XIV. The most important documents for this study are respectively the *Journal de la Communauté de Saint-Louis (1852–1890)* and the *Echo de Saint-Louis* (1906–1920) on the one hand, and the *Bulletin de la Congrégation de Saint-Joseph de Cluny* (1885–1904) and the correspondence exchanged between the administration and the *Soeur Supérieure*, on the other. The missionaries' archives need to be approached as a particular discourse. They considered belief systems other than Christianity as superstition. The researcher needs to sort out what is important and what is trivial. Such a task can be accomplished by comparing the *Journal* with other sources. The missionaries' archives are, nevertheless, valuable because of the quantity of information they contain, as missionaries had the advantage of staying longer in Saint-Louis and having more contacts with the urban dwellers than did the officials. And their archives are of great value to historians if submitted to a critical analysis, and they can be used to supplement the official archives. Despite these shortcomings, these archival materials deal with issues the social and urban historians are concerned with.

Thus, the available evidence has strengths as well as deficiencies and gaps. Only the careful reading of the archives (official and from the missionaries) and the comparison with other sources (government and semi-government publications, secondary literature, and oral data) can provide a global picture of the medical events, processes, and structures analyzed here. One needs to keep in mind the context in which these various written documents were produced. Their authors were influenced in one way or another by the evolutionary theory, which argued that all societies evolved from Savagery to Barbarity to Civilization, from irrationality to rationality, and from superstition to enlightenment. The final stage of evolution was believed to be the Western society. This conviction underlines all the value judgments they made about the indigenous societies. In addition, the authors of the letters and reports in the archives were undergoing a process of culture shock and their opinions about the local population were often characterized by ethnocentrism. This is the reason why their reports are full of racial, class, religious, ideological, and cultural biases. Written during the colonial period, they are characterized by the "civilizing mission" *mentalité* and describe the changes that occurred in terms of transformation from barbarism to civilization.

Although the official records contain the deficiencies and gaps described above, they are, however, of great value to this study. They present detailed information on various aspects of city life, which, if carefully analyzed and compared with other sources, can provide some elements of answers to the research questions raised in the introduction of this study.

Oral sources also provided insight into the historic indigenous conceptions of health, ill health, and health care practices, as well as the reactions of the Senegalese to the outbreaks. However, healer informants were more preoccupied with their credentials, credibility, income, and reputation as healers of great fame than about the indigenous theories of disease. They had personal interests to defend and promote, which affected their testimonies in terms of coherence. The informants from Guet-Ndar were very defensive about their present-day reputation as "rebels" and unsanitary subjects who lived in a "city within the city" and never obeyed any laws. Scholars need to keep in mind that history is part of the fishermen's daily struggle in order to reshape their identity and culture.

*Colonial Pathologies, Environment, and Western Medicine* is divided into six chapters. Chapter 1 discusses the process of growth of the city, the contemporaries' perceptions of the causes of health problems, and the actual factors that contributed to ill health. Chapter 2 examines the elaboration by the French health authorities of "knowledge" about the etiology of yellow fever, yellow fever mortality, and the medical, political, and social responses to it, ranging from quarantines, *cordons sanitaires*, and repatriation to medical profiling, segregation, and protests. Chapter 3 deals with cholera epidemics, which were described as the "Black man's disease," in contrast to yellow fever that was perceived as the 'White man's disease." Chapter 4 examines the conflict of interests and the crisis of confidence that the imposition of an automatic annual quarantine generated not only between various constituencies in Senegal but also between Senegal and neighboring colonies. It also analyzes other drastic steps taken to prevent another outbreak of yellow fever and responses to them, as well as the panic and the disorganization that the 1900 yellow fever epidemic provoked. Chapter 5 analyzes the search by the authorities in Paris and Saint-Louis for a permanent solution to the persistent "native problem" of ill health and to the issue of quarantines that paralyzed the economic life of French West Africa. It deals with the changing paradigm with the advent of the bacteriological revolution. The chapter also discusses the recommendations made by the Mission of Inquiry (February–March 1901), and the adoption of new public health measures or sanitary engineering in the area of vector control. Chapter 6 examines Saint-Louis' experience with bubonic plague, the medical and political responses to the plague epidemic, which were unprecedented in character and scale, as well as the responses and initiatives of the urban poor. The evidence presented in this chapter indicates that despite the new scientific knowledge brought about by the bacteriological revolution, the stigmatization of the urban poor continued to dominate the medical thinking and practice of the colonial and medical authorities.

## Notes

1. See, among others, David Arnold, *Colonizing the Body: State Medicine and Epidemic Disease in Nineteenth-Century India* (Berkeley: University of California Press, 1993); Arnold, *The Tropics and the Traveling Gaze: India, Landscape, and Science, 1800–1856* (Seattle and London: University of Washington Press, 2006).
2. E. Aubert, "Un peu d' hygiene coloniale," *Les Annales coloniales*, no. 11 (Nov. 1, 1901), 92–93.
3. Ibid., 93.
4. For more on colonial SPAS, see Eric T. Jennings, *Curing the Colonizers: Hydrotherapy, Climatology, and French Colonial SPAS* (Durham and London: Duke University Press, 2006.
5. Mary Douglas, *Purity and Danger: An Analysis of the Concepts of Pollution and Taboo* (New York: Routledge, 2002), 4, 9, 44. See also Donald Reid, *Paris Sewers and Sewermen: Realities and Representations* (Cambridge, MA: Harvard University Press, 1991), 2.
6. Michael Taussig, *Shamanism, Colonialism, and the Wild Man* (Chicago: University of Chicago Press, 1987), 5–7.
7. Donald Reid, *Paris Sewers and Sewermen*, 3.
8. Paul M. Hohenberg and Lynn Hollen Lees, *The Making of Urban Europe, 1000–1950* (Cambridge, MA: Harvard University Press, 1985).
9. See W. Christaller, *Central Places in Southern Germany*, trans. C. W. Baskin. Reprint. (Englewood Cliffs, N.J.: Prentice-Hall, 1966). Cited by Hohenberg and Lees, 49–51.
10. Paul H. Hohenberg and Lynn Hollen Lees, *The Making of Urban Europe, 1000–1950*, 4.
11. For more on the Senegalo-Mauritanian zone, see David Robinson, *Paths of Accommodation: Muslim Societies and French Colonial Authorities in Senegal and Mauritania, 1880–1920* (Athens: Ohio University Press, 2000).

12. Immanuel Wallerstein, *The Modern World System I: Capitalist Agriculture and the origins of the European World Economy in the Sixteen Century* (New York: Academic Press, 1974); Wallerstein, *The Modern World System II: Mercantilism and the Consolidation of the European World-Economy, 1600–1750* (New York: Academic Press, 1980); Wallerstein, *The Modern World System III: The Second Era of Great Expansion of the Capitalist World-Economy, 1730s–1840s* (New York: Academic Press, 1984); Wallerstein, "The Itinerary of the World System Analysis, or How to Resist Becoming a Theory," in J. Berger and M. Zelditch Jr. (eds.), *New Directions in Sociological Theory* (Lanham, MD: Rowman & Littlefield Publishers, 2002), 359–376.
13. Jan de Vries, *European Urbanization, 1500–1800* (Cambridge: Harvard University Press, 1984).
14. G. J. Telkamp, *Urban History and European Expansion. A Review of Recent Literature Concerning Colonial Cities and a Preliminary Bibliography* (Leiden Centre for the History of European Expansion, 1978).
15. R. Ross and J. Telkamp, *Colonial Cities. Essays on Urbanism in a Colonial Context* (Dordrecht, Netherlands; Boston: M. Nijhoff, 1985).
16. David Simon, *Cities, Capital and Development. African Cities in the World Economy* (London: Belhaven Press, 1992); C. Coquery-Vidrovitch, "The Process of Urbanization in Africa," *African Studies Review*, vol. 34, no. 1 (April 1991), 1–98; Frederick Cooper, *The Struggle for the City* (Beverly Hill, CA: Sage Publications, 1983), 10; Peter Reeves, "Studying the Asian Port City," in Frank Broeze (ed.), *Brides of the Sea. Port Cities of Asia From the 16th–20th Centuries* (Honolulu: University of Hawaii Press, 1989), 29–53; David Prochaska, *Making Algeria French. Colonialism in Bône, 1870–1920* (New York: Cambridge University Press, 1990), 13; Anthony D. King, *Urbanization, Colonialism, and the World-Economy. Cultural and Spatial Foundations of the World Urban System* (New York: Routledge, 2002); Franklin W. Knight and Peggy K. Liss (eds.), *Atlantic Port-Cities: Economy, Culture, and Society in the Atlantic World, 1650–1850* (Knoxville: University of Tennessee Press, 1991); Phyllis Martin, *Leisure and Society in Colonial Brazaville* (New York: Cambridge University Press, 1995); Paul Tiyambe Zeleza and Cassandra Rachel Veney, *Leisure in Urban Africa* (Trenton, NJ: Africa World Press, 2003); Sarah Nuttall and Achille Mbembe (eds.), *Johannesburg: The Elusive Metropolis* (Durham, NC: Duke University Press, 2008). This list is not exhaustive.
17. P. Alquier, "Saint-Louis du Sénégal sous la Révolution et l'Empire, 1789–1809," *B.C.E.H.S. de l'A.O.F.* (1922); R. Rousseau, "Le site et les origins de Saint-Louis," *La Géographie*, vol. 44, nos. 2–3–4–5 (July, Sept., Nov., Dec. 1925); Roger Pasquier, "Villes du Sénégal au XIXe siècle," *Revue d'Histoire d'Outre-Mer* T. XLVII (1960), 387–426; Camille Camara, *Saint-Louis-du-Sénégal. Evolution d'une ville en milieu Africain* (Dakar: IFAN, 1968); A. S. Diop, "La foundation de Saint-Louis," *Bulletin de l'I.F.A.N.* T. 37.B.2 (1975), 1–50; F. Brigaud and J. Vast, *Saint-Louis du Sénégal, ville aux mille visages* (Dakar: Édit. Clairafrique, 1987).
18. F. Deroure, "La vie quotidienne à Saint-Louis par ses archives, 1779–1809," *Bulletin de l'I.F.A.N.* B, T. 26.3–4 (1964); L. C. Barrows, "The Merchants and General Faidherbe: Aspects of French Expansion in Senegal in the 1850's," *Revue Française d'Histoire d'Outre-Mer*, 223 (1974), 218–234; G. E. Brooks, "The Signares of Saint-Louis and Gorée: Women Entrepreneurs in Eighteenth Century Senegal," in N. J. Hafkin and E. G. Bay (eds.), *Women in Africa. Studies in Social and Economic Change* (Stanford, CA: Stanford U. Press, 1976), 19–44; F. Zuccarelli, "Les traitants des comptoirs du Sénégal au XIXe siècle," *Entreprises et Entrepreneurs en Afrique* (Dec. 1981); François Manchuelle, "Métis et colons: la famille Devès et l'emergence politique des Africains au Sénégal, 1881–1897," *Cahiers d'Études Africaines*

96.XXIV-4 (1984), 477–504; L. C. Barrows, "L'Oeuvre, la carrier du general Faidherbe et les débuts de l'Afrique Noire Française: une analyse critique contemporaine," *Le Mois en Afrique* 235–236 (Aug.–Sept. 1985) and 237–238 (Oct.–Nov. 1985); J.-P. Biondi, Saint-Louis du Sénégal: *Mémoires d'un Métissage* (Paris: Denoel, 1987).
19. D. Bouche, "Les villages de liberté en A.O.F.," *Bulletin de l'I.F.A.N.*, vol. XI, B no. 3–4 (1949), 491–540, and vol. XII, 1 (1950), 135–215; D. Bouche, "L'école française du Sénégal de 1850 à 1920," *Revue Française d'Histoire d'Outre-Mer* 223 (1974), 218–234; F. Zuccarelli, "Les Maires de Saint-Louis et Gorée de 1816 à 1872," *Bulletin de l'I.F.A.N*, B, vol. 35, no. 3 (1973); F. Zuccarelli, *La vie politique sénégalaise, 1789–1940* (Paris: C.H.E.A.M., 1987); M. Diouf, "The French Colonial Policy of Assimilation and the Civility of the Originaires of the Four Communes (Senegal): A Nineteenth Century Globalization Project," *Development and Change* 29.4 (1998), 671–696; H. O. Idowu, "The Establishment of Elective Institutions in Senegal, 1869–1880," *Journal of African History* 9 (1968), 261–77.
20. Alain Sinou, *Comptoirs et villes coloniales du Sénégal: Saint-Louis, Gorée, Dakar* (Paris: Ed. Karthala, 1993).
21. K. D. Patterson, "Disease and Medicine in African History: A Bibliographical Essay," *History in Africa*, 1 (1974), 147.
22. Claude Pulvenis, "Une Epidémie de fièvre jaune à Saint-Louis du Sénégal (1881)," *Bulletin de l'Institut Fondamental de l'Afrique Noire*, B, XXX (1968): 1353–1373; see also Dr. Jean-Marc Clément, "Histoire médicale de Saint-Louis," *Bulletin d'Epidémiologie*, no. 9 (September 1985), 210–212; no. 10 (October 1985), 241–243; no. 11 (November 1985), 274–276; no. 12 (December 1985), 301–304.
23. Myron Echenberg, *Black Death, White Medicine: Bubonic Plague and the Politics of Public Health in Colonial Senegal, 1914–1945* (Oxford: James Curry, 2002), 8.
24. David Robinson, *Paths of Accommodation: Muslim Societies and French Colonial Authorities in Senegal and Mauritania, 1880–1920* (Athens: Ohio U. Press, 2000).
25. Charles Rosenberg, "What is an Epidemic? AIDS in Historical Perspective," *Daedalus*, 118 (1989), 1–17.
26. Michel Foucault, *Naissance de la biopolitique: Cours au Collège de France (1978–1979*, édition établie sous la direction de François Edwald et Alessandro Fontana (Paris: Seuil, 2004);Foucault, *L'Herméneutique du sujet, cours au Collège de France, 1981–1982* (Paris: Gallimard-Seuil-EHESS, 2001); Foucault, *Il faut défendre la société, cours au Collège de France, 1974–1975* (Paris: Gallimard-Seuil-EHESS, 1997); Foucault, *Histoire de la sexualité, tome II: L'Usage des plaisirs*, and tome III: *Le Souci de soi* (Paris: Gallimard, 1984); Foucault, *Histoire de la sexualité, tome I: La Volonté de savoir* (Paris: Gallimard, 1976); Foucault, *Surveiller et punir. Naissance de la prison* (Paris: Gallimard, 1975); Foucault, *L'Archéologie du savoir* (Paris: Gallimard, 1969); Foucault, *Naissance de la clinique. Une archéologie du regard médical* (Paris: P.U.F., 1963; réed. 1972); Foucault, *Folie et déraison. Histoire de la folie à l'âge classique* (Paris: Plon; 1961); rééd. modifiée: *Histoire de la folie à l'âge classique* (Paris: Gallimard, 1972).
27. Alain Corbin, *Le miasme et la jonquille: l'odorat et l'imaginaire social, XVIIIe-XIXe siècles* (Paris: Aubier Montaigne, 1982).
28. Christian Delacroix, Francois Dosse, and Patrick Garcia, *Les Courants historiques en France, XIXe-XXe siècle*. Éd. revue (Paris: Gallimard, 2005); Peter Burke, *The French Historical Revolution: The Annales School, 1929–89* (Stanford, CA: Stanford University Press, 1990); François Dosse, *L'Histoire en miettes: Des "Annales" à la "nouvelle histoire"* (Paris: Éditions La Découverte, 1987).

29. See Michel de Certeau, *La Possession de Loudoun* (Paris: Gallimard, 1970; Julliard, 1990); *La Culture au pluriel* (Paris: Seuil, 1974; 1993); *L'Écriture de l'histoire* (Paris: Gallimard, 1975; 1984); *L'Invention du quotidien*, vol. 1: *Arts de faire* (Paris: Gallimard, 1980; 1990); *L'Invention du quotidien*, vol. 2: *Habiter, cuisiner*. Avec la collaboration de Luce Giard et Pierre Mayor (Paris: Gallimard, 1994); *La Fable mystique*, vol. 1: *XVIe-XVIIe siècle* (Paris: Gallimard, 1982; 1987); *Heterologies: Heterologies: Discourse on the Other*. Trans. by Brian Massumi (Minneapolis: U. of Minnesota Press, 2006).

30. See *Annuaire de la Marine et des Colonies* (Paris: Imprimerie Impériale, 1871).

CHAPTER ONE

# The Making of a Colonial City

This chapter discusses the process of growth of the city, the contemporaries' perceptions of the causes of health problems, and the actual factors that contributed to ill health. It argues that the growth of Saint-Louis and the inadequacies of public health policies led to the inequality in health and life chances among various segments of the urban population. The first section discusses the process of city growth, its geography, economy, society, and government. The second section deals with the city's health problems and medical institutions.

## From a Fishing Village to a Trading Center

Built around Ndar island at the mouth of the Senegal River, Saint-Louis was founded in 1659 as a fortified refreshment and trading station for the French Company du Cap Vert et du Senegal, operating in the context of the emerging French merchant empire, which received monopoly trading concessions.[1] At the beginning, Saint-Louis's spatial layout resembled other European trading posts on the African coast and in the Caribbean which comprised the following elements: a) the company's warehouses where merchandise was stored; b) a fort called *Habitation*, which included the director's residence, the church, and a room for meetings and court sessions; c) several artisans' shops and taverns[2]; d) and the huts built by the Africans, in majority the Wolof Muslims from Walo and Cayor, who lived around the fort. The location of the fort of Saint-Louis was an advantageous position for the defense against an enemy and for commerce, despite the fact that the bar limited the capacity of the port to shelter a large number of ships.

But, early on, the island gained the reputation of being "one of the unhealthiest places on the globe"[3] because of the mortality and morbidity associated with seasonal diseases and allergies. Of course, this statement from Dr. Adolphe Salva was an exaggeration based on local perceptions and not on objective comparative statistics with other port-cities, such as Rio de Janeiro[4] in Brazil or Veracruz[5] in Mexico that had also acquired a bad reputation because of diseases (yellow fever, smallpox, and plague). Rio in fact went through similar programs of disease control and eradication that transformed the city and the country. Saint-Louis's residents had developed a cliché that stood for the common knowledge accumulated over the years of observation, that is to say, "the growth of *baobab* leaves [during the hot and rainy season, *hivernage*] announces the death of the Whites, while their fall [during the dry and cold season] that of the Blacks."[6] Indeed, the cliché reflected the perceived racial susceptibility to seasonal diseases and allergies that would have serious implications for public health policies. The climate and topography were to blame for these illnesses. A huge permanent swamp in the northern part of the island and other swampy areas in the surroundings of Saint-Louis were considered the source of ill health. The rainy or hot season between July and November, although good for navigation and commerce along the Senegal River, brought weather-related ailments; it was "the bad season," the season of the floods and mosquitoes, and "the season of the fleas," during which it was "impossible to close an eye during the night."[7] The rainy season was also the time when mosquito-borne diseases, such as malaria, yellow fever, and dengue fever, often struck the city. The dry or cold season between December and June created severe shortages of fresh drinking water, as the river water became salty around Saint-Louis. A strong wind (*harmattan*) blowing from the east and south between January and March was associated with the proliferation of cases of diarrheal diseases, respiratory infections, eye infection, and flu. Even during the city's peak in the 1890s, the streets leading to the big arm of the river in the northern part of the city were dead ends, obstructed by the accumulation of garbage. The border of the river was also transformed into a dumping ground. In addition, the island on which the trading post was built was sterile and unfit for agriculture, and supply became a permanent concern of the inhabitants of Saint-Louis not only for commercial purposes, but also for their very survival. Their immediate neighborhood[8] supplied them with fish and other products; from the Canaries came vegetables and fruits; the Trarzas Moors at the northwestern end of the Senegalo-Mauritanian zone supplied Saint-Louis with milk and meat, and traders bought their cereal (millet) from the regions of Cayor, Fouta Toro, Galam, Sine and Saloum, and their rice from the *Rivières du Sud*.[9]

Given the nature of these constraints, Saint-Louisians were destined to become "middlemen" in the chain of supply and demand that connected producers and consumers in the Senegalo-Mauritanian zone and France,[10] and survival depended on strategies and policies formulated with great care. Thus, from the beginning, French merchants had doubts about the viability of the trading post. They were only partly right. Indeed, Saint-Louis' experience with sanitation and disease corroborated the merchants' early perceptions. But by the same token, with the acceleration of trade as a result of Colbert's policies, the fort developed into a port-city/capital that became a central place and a major entrepôt in an Atlantic network of trading stations extending north to several French ports (Rouen, Nantes, La Rochelle, Bordeaux and Marseilles), west to the Antilles, Boston and Brazil, and south to West African ports, Gabon, Congo and beyond. The commercial hinterland initially included several seasonal marketplaces (called *escales*) along the Senegal River and its branches.[11]

The successive companies dealt mainly in slaves destined to the Antilles until 1791 and gum. Before the emergence of peanuts in mid-nineteenth century, other products included wax, leather, gold, ivory, and animal skins.[12] But commerce suffered a setback during a period of shifting control over Saint-Louis between the British (1758–1778, 1809–1817) and the French (1778–1809). In the meantime these developments strengthened the control of the business community over the town government. The embryo of administration started in 1764 when the "*Conseil des Habitants*" established the first mayor, Charles Thevenot. In 1782 a direct royal government took over with the appointment of the first governor, Dumontel, and a colonial administration was set up to replace the authority of the directors of the companies. After 1817, with the introduction of free trade, resident French merchants representing Bordeaux commercial firms began to replace merchant directors and employees of concessionary companies and to invest significantly in commerce.[13] In the early nineteenth century the governor was given the title of "Governor of Senegal and Dependencies." It is not until 1840 that the government of Senegal was organized like the other French colonies by an *Ordonnance royale organique*. Its bureaucratic structure developed according to the local needs and the governors' main preoccupations, which remained military until the 1880s.[14]

## Urban Society

Early urban society included the Africans of mainly Wolof origin and the Company agents, who took local women as concubines against the Company's official policy. These temporary unions, called "*marriage à la mode du pays*" (meaning "marriages according to the custom of the land"), promoted some women to the status of *signares*, that is, wealthy entrepreneurs who owned slaves, boats, and houses.[15] Later on, *signares* were chosen because of their status. In any case, these "marriages" were eventually institutionalized and resulted in the second half of the eighteenth century in the emergence of a *métis* community, which was going to play a key role in trade and politics. The growth of trade and the effort to extend the hinterland attracted more people to the trading post, including investors, traders, missionaries and people in search of work, freedom or relief.

The available data indicate that in 1799 the trading post housed 9,000 people, including 600 Europeans, 1,200 *métis*, 3,000 captives, and 4,000 free Africans.[16] At the mid-nineteenth century the population doubled to 14,129 people; by 1900 there were 30,000 residents in the city, as Figure 1.1 indicates. It suggests that the demography of Saint-Louis was characterized by the cycles of growth and decline and that the relative population growth, less than 2% per year after the mid-nineteenth century, reflects the interaction between natural growth, immigration, rural/urban migration, and levels of mortality in the city due to infectious and parasitic diseases, epidemics, natural disasters (floods), the crises of subsistence that accompanied the 1875–1880 economic depression, the decline in the gum trade that accelerated after 1903, and World War I. There are nonetheless two periods of population increase. The first period coincides with the tenure of Governor Léon Faidherbe, who stimulated economic activities through his policy of public works (bridges, docks, and telegraphic lines) and security along the Senegal River in the 1850s–1860s. This period also saw the arrival of the first disaster victims and refugees from the interior to Saint-Louis. Indeed, drought and crop failures in 1863, locust invasion in 1864, in addition to the scarcity of gum, the ongoing wars of conquest against Lat Dior Diop in Cayor and Saloum, and the subsequent famine pushed the country people off the land and drove them to Saint-Louis in search of food, shelter and work.[17] The refugees became part of the "fluctuating population" estimated at 1,182 people in 1877.[18] This rural exodus stimulated the administration, city government, clergy and community leaders to set up the structures of poor relief that became the standard model of intervention used later on to alleviate the suffering of the victims of other calamities, such as floods, fire, and poor harvests. The second period of population increase was in the 1890s with an influx of more European families and troops in Saint-Louis when the city achieved its zenith. It was then the capital of "Senegal and dependencies" (1854–1902), the seat of the Government General of French West Africa (1895–1902), the capital of Senegal (1904–1960), the capital of Mauritania (1904–1957), the headquarters of the naval station, and the point of departure of missions of exploration and military expeditions in the hinterland. Despite the slow population growth, Saint-Louis remained the largest city in Senegal until 1914, when the economic crisis following the outbreak of World War I resulted in the migration of

## Figure 1.1. Saint-Louis Population Growth, 1787–1921

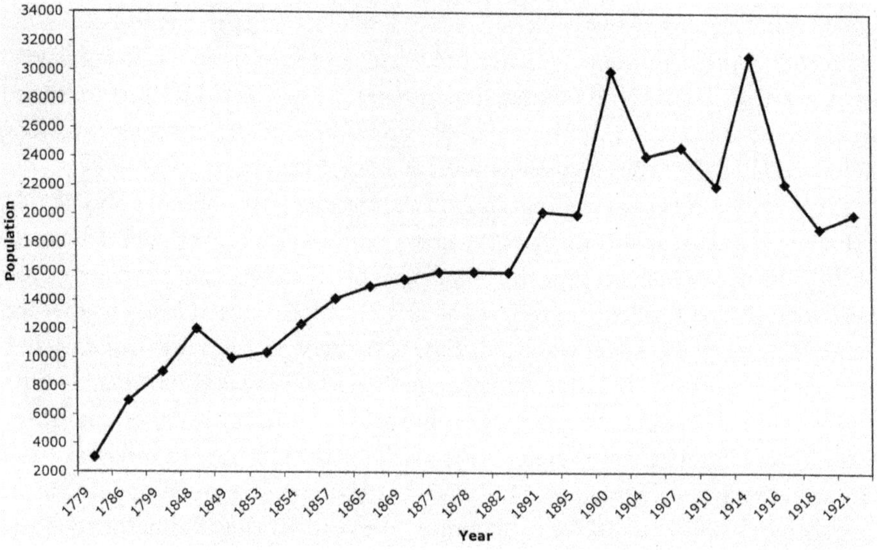

Sources: A.N.S. 22 G 6, Recensement de la Population du Sénégal, 1847–1884, especially pieces 17, 18, 26, 39; 22 G 31 Statistiques Générales, especially pieces 14, 17, 36; Statistiques Générales; Archives Municipales de Saint-Louis, 1 Q 80, Recensement de la population, 1914; Bonnardel, R., *Saint-Louis-du- Sénégal: Mort ou Naissance?* (Paris: L'Harmattan, 1992), 169.

people from Saint-Louis to Dakar, the towns along the railroad and the Senegal River. The population decreased from 28,782 to 22,169 residents in 1916.

The population was racially, ethnically, socially, and culturally heterogeneous, as Table 1.1 shows. The data are scant, however, concerning the distribution of the population according to race, sex, and residence. It was not until 1916 that the censuses on local population included ethnicity. The French, mostly from Bordeaux and Marseille, formed a sociological minority who controlled the colonial government, the public bureaucracy, the army, and the economic sector as *négociants* (wholesalers) and *marchands* (semi-wholesalers or retailers). The *négociants* carried stocks, delivered products, supplied the government with various commodities, and granted credit to their customers. This was the case for Maurel and Prom, Devès and Chaumet, Buhan and Teissère, Delmas and Clastres, Peyrissac and Cie, and Chavanel, among others, who made fortunes trading in gum and millet. The *marchands* were semi-wholesalers and retailers, largely French, who owned their stores. They invested in gum trade, and some even created their own firms.[19] After the 1840s peanuts became the main product.

The *métis* were the mulatto branches of the French families; the Devès, Descemet, Carpot, Valantin, Crespin, Guillabert, d'Erneville, Thévenot, Pellegrin, and

Patterson families are good examples.[20] During the period of Saint-Louis's greatness, the *métis* controlled the deputyship, the city government, and the General Council. Some were *marchands* or semi-wholesalers and retailers; others were in the professions. There were few lawyers. The *métis*, for example, saw themselves as cultural brokers, working for the spread of *l'idée francaise* in the context of an overwhelmingly Muslim culture.[21] The elite group also distinguished itself from the commoners by various nonmaterial symbols related to their style of life: manners and etiquette, titles, membership in the Catholic Church or freemason lodges, marriage with "good" families, a special hairstyle, and pride in having one's ancestors' tombstones in the Catholic cemetery in Sor.[22] M. Marcson argued that there were 30 or 40 powerful families between 1758 and 1854.[23] Various import stores, café-restaurants, and bookstores in Saint-Louis supplied the elite group with consumer goods, ranging from Camembert cheese to Chateau Saint-André wine, that were not accessible to the commoners. Both the French and the rich *métis* formed a status group[24] and considered themselves the "*Grands Hommes*" (big men) in so-

Table 1.1. Ethnic/Racial Distribution of the Population (no Troops or Navy)

|  | 1799 | 1857 | 1869 | 1878 | 1893 | 1895 | 1900 | 1914 | 1918 |
|---|---|---|---|---|---|---|---|---|---|
| FRENCH | 600 |  | 1,000 | 1,300 |  | 2,000 | 1,500 | 2,280 |  |
| CREOLES | 1,200 |  | 2,000 |  |  |  |  |  |  |
| WOLOF | 4,000 |  | 12,000 |  |  |  |  |  |  |
| BAMANA |  | 250 |  |  |  |  |  |  |  |
| PULAAR |  |  |  |  |  |  |  |  |  |
| SONINKE |  | 200 |  |  |  |  |  |  |  |
| MOORS |  |  |  |  |  |  |  |  |  |
| NEW MIGRANTS |  | 1,500 | 450 |  |  |  |  |  |  |
| CAPTIVES | 3,000 |  |  |  |  |  |  |  |  |
| CITIZENS |  |  |  |  |  |  |  | 21,950 |  |
| TOTAL | 9,000 | 12,081 | 15,480 | 16,000 | 20,173 | 20,000 | 30,000 | 28,782 | 19,000 |

Sources: A.N.S. 22 G 6, Recensement de la population du Sénégal, 1847–1884, especially pieces 17, 18, 26, 39; 22 G 31 Statistiques générales, esp. pieces 14, 17, 36; Bérenger-Féraud, De la fièvre jaune au Sénégal (1874), 20–21; Statistiques générales; Archives Municipales de Saint-Louis, 1 Q 80, Recensement de la population, 1914; F. Zuccarreli, *La Vie Politique Sénégalaise (1789–1940)* (1987), 12, 35 ; Régine Bonnardel, *Saint-Louis Du Sénégal : mort ou naissance ?*(1992), 169.

ciety, at the top of a hierarchy founded on wealth, power, and prestige. The objective symbols of status included the ownership of a well-equipped two-storey house, clothing, consumer goods, a degree from France, and having children receiving their education in France.

But in the 1840s and 1850s, as the mercantile empire made way for a free-trade empire, the role of the *métis* as "middlemen" began to decline because of the suppression of the slave trade and of slavery, the decline in the gum trade, the development of the peanut production and the increasing number of competitors (firms, *traitants*, and individual merchants), and the accumulation of debt and subsequent bankruptcy.[25] Instead, domination of the economic as well as the political life by the French *négociants* grew. French merchants, companies and firms seized new opportunities for investment, business and financial dealings, such as renting houses to the government,[26] supplying troops and government employees with millet and rice, and building materials used to pave the city streets and to build the sidewalks. So in the second half of the mid-nineteenth century the *métis* population extended all the way from the elite down to day workers. In the middle, one could find the retail merchants, small shopkeepers, quasi-professionals (teachers and clergymen), civil servants, and saloonkeepers. At the bottom were some *métis* who had little formal education.

The key to a child *métis'* life chances and success depended upon the official recognition by its father, which granted it French citizenship and educational and job opportunities that were not available to those less fortunate who were considered "natural children," labeled *indigènes* or simple French subjects, and stigmatized for lacking a "race," a father, and a family.[27] The advent of legislative and General Council elections in the 1870s would lead to a merciless rivalry between *métis* and "Bordelais," and a tension between the merchant community and governors.

Despite the economic changes, the *métis* did not lose their social and political status. They still had decisive political influence as "*Grands Hommes*" and great prestige. Their names continued to appear on the annual list of "notables" and, under the leadership of the Devès, Carpot and Descemet families, they held government office of one sort or another, particularly in the deputyship until 1914, the municipality as mayors until 1916, and the General Council until 1920. One of them held a high rank in the army: General Amadée Dodds, chief of the French expedition in Senegal and Dahomey, known as the "*vainqueur de Dahomey*." By 1890, however, a new group of merchants, the Libano-Syrians, established in Saint-Louis and engaged in competition with the *métis* as middlemen between *négociants* and Africans in rural areas.

The Wolof formed the majority of Saint-Louisians. They were divided into "higher" and "lower" "castes" or occupational traditions. The most successful Wolofs were the Muslim clerics (*marabouts*), the merchant families (*traitants* and *sous*-

*traitants*) and the bureaucrats (interpreters and other clerks). The leading Muslim clerics were the leaders of the Muslim community and some were allies of the administration. The *traitants*, or retailers, received credit in merchandise from the *négociants* in Saint-Louis and engaged in personalized transactions in the *escales* mainly involving gum. Others were managers on commission or employees on salary. *Traitants* also acted as brokers providing information about what buyers needed in the hinterland and what supplies were available in Saint-Louis. In the late nineteenth century successful *traitants* acceded to the status of "notables" or "esteemed families." The career of Hamet Gora Diop (1846–1910), analyzed by Mamadou Diouf, is a good case in point,[28] even if these cases represented the few people who were engaged on individual trajectories running counter to the collective decline of the members of their class fraction. There were also some independent *traitants* dealing in grain (millet and rice); they had low status compared to the *traitants* in gum.[29] Some Wolofs were sailors (*laptots*) who worked for the French commercial and military fleet on the Senegal River. The leading Muslim merchant families included the Gueye, the Mbengue, the Lo, and the Diop families.[30]

Both the *métis* and Muslim merchants defined themselves as "*enfants du pays*" or "*enfants de Ndar*" (sons and daughters of Ndar) in the sense of "first-class citizens," while others regarded them as individuals from "esteemed families."[31] Thus, selected aspects of the formation of Saint-Louis's merchant elite reveal that not only "marriage," as described above, but also commissions and credit contributed to the building of the traders' careers, especially *marchands*, and *traitants*.

The Wolof constituted most of the "popular classes" in Saint-Louis, to borrow Louis Chevalier's expression.[32] Some were involved in the informal sector. Others were fishermen, day laborers, household servants, peddlers, guides, interpreters, street vendors, or water carriers. The members of "lower castes" were craftsmen (tailors, jewelers, blacksmiths, barbers, shoemakers, weavers, cobblers, potters, etc.), and entertainers (*griots* and *griottes*). Women were rented out as pounders of millet (*pileuses*) and washers (*blanchisseuses*). Many were former slaves. The poor had little formal education. For a long time, the Wolof opposed sending their children to the French school because they feared that their children would become Christians and would lose their cultural (Muslim) identity.[33]

Other ethnic groups came to Saint-Louis either as slaves, as traders, or as migrants in search of work, relief or asylum. The most successful among them became interpreters in Saint-Louis or worked as artisans. The Moors from the Mauritanian zone operated as long-distance traders in the Senegalo-Mauritanian zone. The Bamana (Bambara) first arrived in Saint-Louis as slaves; later, those who migrated to Saint-Louis were avoiding slavery in their homeland in Kaarta and Segu. They settled in Sor and, in order to meet their needs during the rainy season, they cultivated a piece of land granted to them by Jean Beziat, a local merchant. During the dry season, some sought temporary jobs in the city, while others were involved in the informal economy, making charcoal and gardening. As the Bamana were practitioners of African traditional religions, missionaries viewed them as natural allies; their very lifestyle made them "declared enemies" of Muslims: they drank alcohol, ate pork, made *dolo* (indigenous beer or alcohol), and disliked the *marabouts*. In the 1880s some Bamana families converted to the newly introduced Protestantism, considered by the mainstream elite the "Church of the uprooted and helpless people." That was the beginning of a harsh competition between Catholics and Protestants for the Bamana souls. In order to attract the Bamana to Catholicism, the Catholic Church offered free housing to the first seven families of converts, translated the Bible and Catechism into Bamana language, and wrote the Bamana grammar and dictionary.[34] Besides the Bamana, there were other recent rural immigrants from various origins who were selling religious services as healers, diviners, and snake charmers, to name a few. The smaller groups included the Pulaar and the Soninke, who shared the same occupations.

## The Built Environment

Saint-Louis developed on Ndar Island, separated from the mainland by two branches of the Senegal River. A central place divided the post into two sections, the North and the South, a distinction stemming from property or capital ownership. The city extended outside the island with the creation of new quarters. On the Langue de Barbarie, a long portion of land separating the Atlantic Ocean from the small branch of the Senegal River, two settlements developed: Guet-Ndar and Ndar-Toute. Guet-Ndar is the oldest Wolof settlement in the area. It was mentioned on the explorer John Lindsay's 1758 map as a small village with a few huts. By the 1880s Guet-Ndar had developed in disorder in a southward movement, pushing the Muslim cemetery further south. The intervention of the authorities was limited to the opening of one main street in the middle of this big village.[35] In 1837 the administration erected the peri-urban village of Saint-Philippe on the territory of Cayor for captives rescued in the sea after the abolition of the slave trade; a second village, called Bouetville, was created in 1852. The settlement was eventually enlarged in 1864.[36] Ndar-Toute was erected in 1856 on an empty lot north of Guet-Ndar near a place where some families and a *village de liberté* for freed slaves was built in 1848. It developed according to a plan approved by Governor Léon Faidherbe. Saint-Louis also controlled three islets bought by Governor Blanchot in 1700: Babagueye, Safal, and Guebel.[37] In 1884 a new African quarter, Gokhoumbaye, located on the northeastern part of Ndar-Toute, was annexed to the city.[38]

The evidence shows that in the 1840s the built environment comprised 1,569 dwellings, including 315 houses, 68 stores, and 1,186 huts in bricks.[39] A decade later, the dwellings numbered at 3,000, including 200 permanent houses made of brick.[40] A map drawn in 1854 shows that at least half of the island was already occupied. The city did not develop in a spontaneous way but according to a predesigned master plan established by the decree of March 31, 1829, reviewed and extended in 1843.[41] The center of the island was occupied by the government and administrative buildings. Early on, the *métis*, African *traitants*, the Whites "*en ménage*" with *signares*, and their domestic servants of Bamana origin lived in large two-storey houses in the southern part of the city, called *Kertian* (meaning Christian) because of the presence of the cathedral and its Catholic population. Some poor families also lived in the center of the city in 200 to 300 huts and in 30 huts west of the Church Place, huts that did not protect them from the cold weather, and there were regular reports of fire that destroyed hundreds of huts. The southern quarter also included, in addition to the residences of the elite, the military hospital, the civilian hospital, and the artillery.

Fires starting in thatch-roofed dwellings during the cold season (between December and May) were a great cause of concern for the authorities since they resulted in loss of property. Such dwellings increased from 1,200 in 1845 to reach 5,000 in 1857 on the city island. In early January 1857, more than 700 huts caught fire in a couple of days, destroying in passing the ships in construction and boats anchoring in the port of Saint-Louis and causing serious concern to the authorities. Imprudence was one cause of fire but there were other causes as well, chief among them being 'the wind from the east (blowing) without interruption for six weeks.' An eyewitness explained that 'from 31 December to 2 January, the wind had such an extraordinary violence that the blanket of sand that it brought with it completely obscured the sun.' The cold wind dried up the dwellings and made them vulnerable to fire. As a solution, the authorities recommended the prohibition to build thatch-roofed dwellings or the displacement of the surplus population to Ndar-Toute but realized that such solutions were unworkable because not only the urban poor could not afford the cost of building in durable materials but also 'they are attached to their quarter, their neighbors, (and) their extended families.' In addition, 'they can save their belongings, their gris-gris, (and) their wives' jewelry.'[42] These rationalizations help explain the reasons why the policy of accommodation adopted by the administration vis-à-vis the urban poor were slowly replaced by a policy of forced removal of the target group from the city center if they did not meet the requirements of building code. Later on, the administration would use the outbreaks of epidemic diseases as a pretext for accentuating the forced removal of the urban poor from the city center, as the following chapter will show.

In 1859 and 1865 Governor Faidherbe, in the framework of his public works policy—one component of which was called "*la bataille de la paillotte*" or "war on shacks"—cleared the city center between Ribet and Saint-Jean Streets of hundreds of huts. His successor, Valière, continued the same policy and built sidewalks and drainage in the city center and along the docks. After that, the Center and the South became the best places to live in Saint-Louis. In the northern part one could also find the City Hall, the Chamber of Commerce, the firms and banks, the Court and the garrison. The ordinary people, in majority Wolof who came from the continent, were concentrated in the extreme southern part of the city next to the European cemetery, in the northern quarter where the Great Mosque was built by Governor Bouet-Williamez in 1847, and in Guet-Ndar. Some families were authorized to settle in Ndar-Toute in 1848. Freed slaves lived in two "*Villages de Liberté*" built in Sor respectively in 1837 and 1852. The Bamana and the Pulaar lived in separate villages in Sor, which also attracted some wealthy families who built summer houses there.

Fires continued to pose a serious threat to the built environment. Indeed, in 1876, half of the huts in Guet-Ndar went into flames and families became homeless and without resources, making them vulnerable to the activities of such opportunist

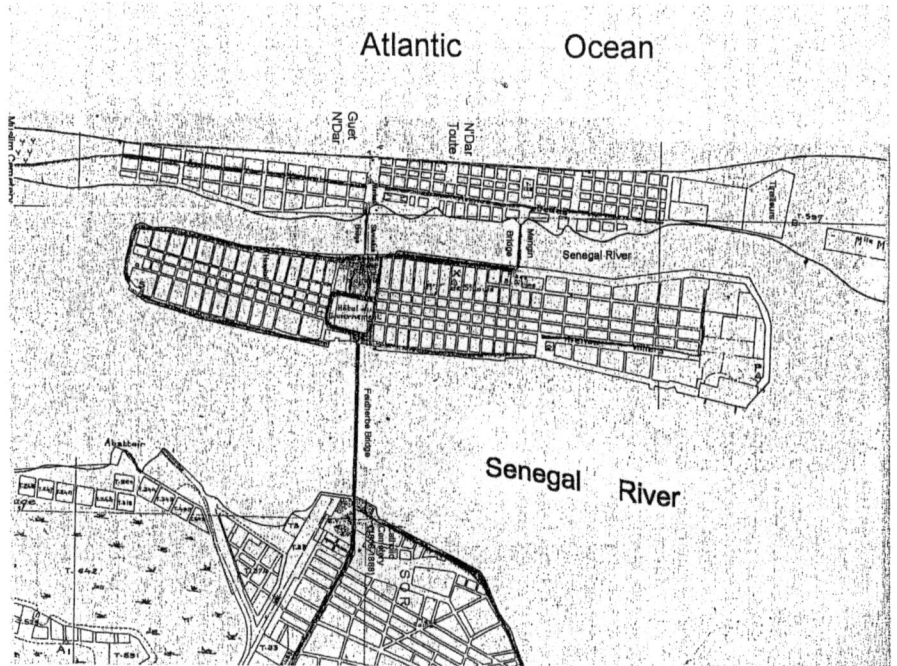

Map 2. Saint-Louis (adapted from Le Genie civil, September 1912, 434)

witch-finders as the Laobe from Cayor, former Tirailleurs soldiers, and other "smart laptots," who explained all misfortunes—including fires—in terms of witchcraft.[43] It should be emphasized that fire also put a strain in race relations. By 1891 tension erupted between the urban poor and the European Spahi troops used as firefighters, as the latter committed acts of violence against and injured the former or insulted them.[44] In 1897 a fire "almost destroyed the totality of the city's *faubourgs*," and led to the issuance, by Mayor Louis Descemet, of fire regulations that extended the zone within which it was prohibited to build huts. It included, on the city island, the area between a street without a name next to Brue Street in the North and Méchin Street in the South; in Ndar-Toute the zone between Servatius Drive and Dodds Street, on the one hand, and the dispensary place, on the other; in Bouetville, the zone between Gopilleau La Roque and Tour de Sor; and in Guet-Ndar, the zone between Servatius Drive and the Batterie Place. The ordinance had provisions for street alignments and enlargements [to six meters], and a distance of six meters between huts and dwellings in wood and masonry. The urban poor were given an authorization for a temporary occupation of their existing dwellings and a three-month deadline from 1 December to rebuild their houses with non-inflammable materials. Violators were threatened with severe sanctions, including the destruction of their property at their own expense.[45]

## Tensions and Divisions

The expansion of Saint-Louis did not take place without tensions. Early on, the main source of stress originated from an urban society itself divided according to wealth, power, and legal status. Indeed, the division of the built environment in the mid-nineteenth century between a "Center," a "South," and a "North," on the one hand, and between the city island and the slum neighborhoods (*faubourgs*) on the continent, on the other, reflected the division of the society between the rich and the poor. Two contemporaries, Frédéric Carrère and Paul Holle, observed that there was a tension between the "Southerner" and "Northerner" city dwellers.

> People from the South made a show of superiority over people from the North based on their claim of the right of the first inhabitants; frequent conflicts occurred between the two quarters and the battles took place near the City Hall during which the adversaries, women in particular, did not hesitate to make loud outcries and use of punches.[46]

The spatial layout of Saint-Louis came to reflect its historical evolution and the occupational structure. Politically, the center, with its government buildings, the governor's residence, and garrison, symbolized power and authority. Socially, the spatial layout of the city expressed social divisions and status, with the notables in the South and the ordinary people in the North. Culturally, it "stored" sentiment and expressed identity: the South was perceived as "Christian," while the North was considered "Muslim."[47] Later, after the government undertook to drain swamps, bogs, moats, and other sites of standing water in the northern part of the island, the distinction between the South, Center and North slowly subsided because some Christian families and businesses settled in the North of the city as well. However, the poor quarters continued to be considered an "uncivilized" space or an underworld inhabited by people perceived as outcasts. The pathological character of the urban existence and the "sanitation syndrome" that characterized the colonial mentality contributed to the identification of the urban poor quarters with disease and to defining their residents as carriers of germs and "dangerous."

There is evidence that there were tensions within the dominant class that were perceptible in the associational life of the elite in various organizations. The most influential but also controversial association was the Freemasonry, the *Grand Orient de France*, whose lodge in Saint-Louis, *L'Union Sénégalaise*, included both Whites and *métis*. It was dissolved in 1876 by Governor Valière after a dispute erupted between Catholics and Freemasons about the "proper" burial ceremony to organize for M. Pallegry, Head of *Ponts et Chaussées* Service, who was claimed by both parties as their own member. The association was re-authorized in 1880 and, by 1885, one of its members, Auguste Foret, founded a weekly newspaper, called *Le Réveil du Sénégal*. He used it as a platform to criticize the government policies, especially the failure of anti-yellow fever measures, and the high society, and to wage propaganda warfare against the Catholic Church and its members. But the shareholders withdrew their support after a series of successful lawsuits against A. Foret, who was forced to shut it down in 1887 after a trial, a fifteen-day jail sentence and the payment of fines. Foret's enemies described him as a "traitor" who betrayed the very foundational principles of the government of the Republic, a "rebel" whose imagination knew no limits and whose hatred and jealousy guided his attacks against "well-intentioned colonial administrators, the most honest magistrates, the governor, and the leading Senegalese families," in the words of the elite women who petitioned the governor.[48] The freemasons and their allies also met in another organization, called the *Cercle de la Concorde*, which organized receptions and concerts that made Saint-Louis a "city of joy," a "jubilant city."[49] However, the Freemasons' influence and their anti-clerical propaganda eventually led to the dismissal of the Catholic nuns from the Civilian Hospital in 1905 and the laicization of education. The Catholics had their own organizations, such as the *Cercle Catholique*, inaugurated in 1897 and whose main goal was to fight the anti-clericalism of the Freemasons.

Even the *métis* did not form a monolithic bloc. The activities of the "*Cercle des Habitants Notables du Pays.*" between 1887 and 1888, were plagued with "disorders" and "complaints."[50] G. W. Johnson argued that, toward the end of the century, there were within the *métis* community family rivalries (the Carpot/Descemet "clan" vs. the Devès "clan"), geographic rivalries (Saint-Louis vs. Gorée/Dakar/Rufisque), occupational distrust (public vs. private sectors, urban vs. rural occupation), and divergent political interests (pro-Bordelais vs. pro-local businesses vs. pro-Africans).[51]

The division between citizens and subjects, or popular classes, was another source of tension within the social system. The naturalization of Saint-Louis's residents as French citizens in 1833 introduced social status as a new criterion for social mobility and put in motion the competition between Saint-Louisians and new immigrants from other towns and villages in the countryside, now called "*niakh*" or "bush people." The 1914 census indicates that in Saint-Louis there were 21,950 French citizens (including many Muslims), 6,766 French subjects (non-citizens), and 66 foreign subjects.[52] In 1916 the law of "four communes" made the Africans born in the communes of Dakar, Gorée, Rufisque, and Saint-Louis French citizens. The struggle for citizenship prompted pregnant women from the countryside to go to Saint-Louis for child birth so that their children could become French citizens. In any case, the tension within the social system erupted into serious social and political dispute during the campaign for parliamentary elections in 1914 when "new men," essentially Blacks, aspiring to the privilege of political participation, rose in society through their personal ability and ambition and with the support of *métis* sympathizers, used the vote as an instrument of emancipation from the French elite and conservative *métis*, and added new families to the list of "notables." Chief among them was Blaise Diagne, who was elected deputy in 1914, defeating F. Carpot, L. Pellegrin, G. Crespin, and Heinburger. The same year, Galandou Diouf, Bakar Waly, and Ahmed Khoury Sene (Ameth Fall) became members of the General Council.[53] Pierre Chimère became the first Black mayor in 1916 with the support of the former *métis* mayor Descemet. But these "new men" did not emerge all at once. There is evidence that as early as 1892 a group of Blacks created the "*Cercle des Bons Amis*" and organized regular meetings in a building located on Saint-Joseph Street owned by Pedre-Alassane. The founding members of the *Cercle* were Abdoul-Karim, Amadou-Cissé, Ousmane-Sarr, and Auguste Lame.[54] Another association, called the "*Cercle du Progrès*," was created in 1893 on Governor Lamothe's initiative, also met regularly in Ndar-Toute, including Cire-Diop, Masseck-Seck, Diam Ngo, Sadiambane, Bakry-Gueye and Mamane.[55] By 1900 a *Comité Indigène* was building support among the French and *métis* sympathizers for an inclusion of Africans on their electoral lists. Twelve years later, some Wolof civil servants used *l'Aurore*, a cultural association founded by a Koranic schoolteacher, Papa Mar Diop, as an outlet for the expression of their political views.[56]

The introduction of Blacks into the political arena, even if it did not significantly affect the health policies and priorities of the colonial administration, nonetheless provided the French colonial authorities, engaged in what David Robinson called a "path of accommodation,"[57] with additional mediators who could convince the Africans to accept French medicine, especially the sanitary measures adopted during epidemic crises.

## Climate, Filth, and Disease

The growth of Saint-Louis and the unequal distribution of wealth and resources that accompanied it were an important cause of ill health. Although early reports romanticized soldiers dying from "nostalgia,"[58] the general living conditions in Saint-Louis were poor in terms of housing, health, and diet, especially for the working class and the poor. Pierre Loti, a Navy officer and writer, described Guet-Ndar in 1873–1874 as overcrowded, ill-ventilated, unprovided with privies, and with no social services.[59] The children playing in the street had "big bellies," which is a sign of malnourishment. On various occasions, the administration complained that there were too many beggars in the city, including "a great number of strangers."

In addition to periodic fires, "fevers" and bad odors from wastes were chronic problems of the first order during the dry and cold season. Indeed, there were reports of the existence in the city of uncollected wastes (putrefying garbage, manure, human excrement, carcasses of dead animals, and kitchen garbage), areas of standing water, and lack of fresh drinking water. Horses, mules and camels used for commercial transportation left pounds of manure and gallons of urine daily on the streets. Stench coming from stables operated by the cavalry was unbearable. Sheep and goats often ran loose in the city streets. There were reports of cows attacking people in the city streets. Although some people were prosecuted from time to time for breaking the ordinances on cleanliness and free movement of animals, such laws were not rigorously applied. Ships, trains, and caravans involved in international trade brought infectious and parasitic diseases to the city, eleven of which were on the watch list. Among them, yellow fever, cholera, and bubonic plague were the most feared because they were spectacular in their signs and symptoms, while malaria, which killed more people than these diseases year in, year out, was viewed as a natural fact of life because it was a silent killer.

Houses and businesses often lacked adequate ventilation. Overcrowded slum areas at the extreme North and South of the city and in the slums facilitated the spread of infectious agents. A law was passed in France in 1850 against unsanitary houses, but enforcement did not begin for three decades. Overcrowding also affected garrisons that had insufficient ventilation and little sunlight. The Military

Hospital and the Civilian Hospital that were not equipped with separate wards for the sick and the convalescents were responsible for the spread of many infectious diseases. These living conditions, in turn, shaped people's life chances and mortality rates.

It is difficult, however, to establish a positive correlation between living conditions and disease based on the statistics of hospital admissions and even on mortality rates. The surviving hospital records since 1889 and the Annual Reports sent to the governor-general since 1896 provide an incomplete picture of the trends in the incidence in Saint-Louis of "silent" killers that contributed to morbidity and mortality for two reasons. First, the official classification of diseases and allergies during most of the nineteenth century was very problematic because these conditions were vaguely understood and ill-defined. For example, a wide variety of conditions were treated under the disease category of "fevers."[60] The "dysentery" category included all types of bowel infections. The sexually transmitted diseases were reported under two different labels: 1) venereal diseases, including syphilis and gonorrhea, and 2) infections of the urinary tract. Second, the low admission rate compared to the size of the urban population, as well as the type of diseases treated, are certainly not a reflection of the actual situation. Instead, they result from a combination of factors, including the unequal access to health care for the majority of residents and the limits of French medicine itself. However, despite the limitations of official statistics, it is possible to identify from various sources the factors that influenced the (re-)emergence and spread of diseases and to examine the trends in incidence and prevalence of major diseases in the city by combining various sources.

*Hivernage*-related diseases were the leading causes of ill health in Saint-Louis. Early on, French health officials made a connection between the rainy season and an increased exposure to and increased number of allergies and vector-borne diseases. Dr. Berenger-Féraud, who worked in Saint-Louis as Chief Medical Officer from 1873 until 1874, observed that, during *hivernage*, the combination of heat, thirst, fatigue, bad dreams, and mosquitoes seriously compromised the health of city dwellers. He gave a long list of hot weather–related ailments: fevers, hepatitis, skin diseases, boils, respiratory complaints, diarrhea and dysentery, rheumatism, and eye infections (for those who spent their nights outside in the open). The evidence shows that the majority of patients in those years were admitted to the hospital during the *hivernage* months, especially in June, July and August, suggesting a normal seasonal pattern of infectious and parasitic diseases. The three main *hivernage*-related diseases were yellow fever (chapter 2), malaria, and dengue fever. Malaria was responsible for great mortality and morbidity in Saint-Louis. Actually, it killed more people than the much-feared yellow fever, and it was a major threat especially to children who had not developed immunity to fight infection yet. It is caused by *Plasmodia falciparum* parasites transmitted by the bite of infected female

*Anopheles* mosquitoes. The case rate increased sharply in July and reached its peak between August and October before declining in November and December. Malaria became endemic in West Africa. The situation remained so despite the clean-up campaigns and an effort at public education about the role of the mosquitoes in the transmission of the disease undertaken after 1900. More Europeans were admitted to the hospital for malaria than the *indigènes* (the urban poor and working class). It would be misleading to conclude, however, that Europeans were more affected by malaria than the *indigènes*. The high admission rates of the Europeans have to do with the differential access to health care and with the fact that the majority of the persons who became ill were received at the hospital as outpatients and the majority of cases went undiagnosed. The incidence of malaria among the European population in 1910 prompted Governor Jules Peuvergne to enforce the regulation of 23 February 1905, fixing two years as the duration of the effective stay of the civil servants in Senegal.[61] But many civil servants stayed in the colony well beyond their term limits despite the conventional wisdom that believed the climate of Senegal was "debilitating and unfavorable to Europeans."[62] The environmental paradigm was used to simplify a complex problem.

Dengue fever was another serious cause of ill health during the *hivernage*. Its virus, as in the case of yellow fever, was transmitted by the bite of infected *Aedes aegypti* and *Aedes Albopictus* mosquitoes that were breeding in stagnant water and discarded water containers. The mosquitoes, vectors of dengue, were also associated with shipping and harbors. Dengue was a disease typical of children but it affected adults as well, and physicians often confused dengue fever with other fevers, making it difficult to obtain clear statistics. Official records contain occasional references to dengue outbreaks that the medical officers viewed basically as a problem of public and domestic sanitation. The 1889 and 1890 *hivernage* seasons witnessed two virulent dengue outbreaks in Saint-Louis that infected about 1,000 Europeans and thousands of *indigènes*, disrupted all the public services and businesses, and probably involved some complicated cases that had fatal outcomes. Population growth, uncontrolled urbanization, and the movement of people into new areas created new breeding grounds of mosquitoes (empty cans, discarded pots and pans, broken bottles, watering places without adequate drainage) and contributed to the periodic dengue outbreaks.[63] So weather-related diseases and the environmental contamination of the city-island itself constituted a threat to the health of the residents.

Problems linked to drinking and bathing in polluted water and to sewage disposal were the second major causes of ill health in Saint-Louis. The Senegal River water around Saint-Louis became fresh in mid-July and could be used safely through December; from January on, the river water turned brackish and was no longer drinkable. The water supply problems were responsible for the high incidence of diarrhea, dysentery and other gastrointestinal disorders, and typhoid fever during the dry and cold season (between December and mid-July).[64] Two cholera epidemics in 1868 and 1869 claimed the lives of thousands of residents, as chapter 3 shows.[65] January 4 was the official date for the distribution of water from government cisterns to civil servants and troops, the civilian hospital, the Roignat garrisons, and the *Genie* workshops.[66] Since only 100 city dwellers had private cisterns, the remaining residents had to rely on contaminated river water and on wells that they dug in the dunes in Sor or in the Pointe de Barbarie. In 1862 the Inspector of the *Service des Ponts et Chaussées* acknowledged that "the lack of water is one of the principal obstacles to the extension of our domination." Even the completion of the water supply project in 1886 provided pure water only between January and April to some sections of the city. Saint-Louis' expanding population in the 1890s increased the demand for water and led to more waste matter being discharged directly into the river and to more exposure to disease.

Beside the diseases of the intestinal system, the cold season was also responsible for a variety of respiratory conditions from the bronchitis/pneumonia/influenza group, labeled as "sporadic" diseases.[67]

Smallpox, a virus causing *variola major* and *variola minor* in humans, was the third serious threat to life in Saint-Louis. Like yellow fever and cholera, smallpox also spread throughout the world with the intensification of trade. The appearance of skin lesions two weeks after infection, and possibly death a week or two after that made it a deadly threat to health. Also feared was the long convalescence the survivors went through and the pock-marks, especially on the face, that gave them great anguish. Highly contagious and killing 20% of those infected, the virus could survive in clothes and bedding and be transmitted by travelers (soldiers, pilgrims, and traders). Inoculation or "*variolation*"—that is, the transfer of smallpox matter from infected persons to the non-immune susceptible through an incision in the arm—was practiced before the arrival of the Europeans to protect uninfected people against smallpox infection. The French introduced smallpox vaccinations that could provide lifelong immunity but the poor quality of vaccine shipped from Bordeaux and the reluctance of the population to seek vaccination and re-vaccination contributed to the spread of smallpox in the city. Smallpox affected mainly the urban poor and the working classes, especially the children and migrants new to the city. The first major smallpox outbreak in Saint-Louis occurred in 1864, when drought and famine in the hinterland forced thousands of country people to seek refuge in Saint-Louis. The concentration of people in the city provided an important reservoir of non-immune susceptible and an opportunity for the virus to become endemic along the railroad Dakar-Saint-Louis. The most serious outbreak of smallpox occurred in 1888–1889, when there was an influx of Europeans and troops in Saint-Louis. Officially, it claimed thirty lives but the real figures may have been higher in Sor, Ndar-Toute, and Guet-Ndar.[68] The following table provides a synopsis of diseases and other challenges that had a direct impact on the city's health.

Sexually transmitted diseases were the fourth major cause of ill health in Saint-Louis. Syphilis was the most important infection among the *indigènes* who sought medical attention from the French physicians and nurses between 1889 and 1898; it was followed by "surgical" diseases (wounds, burns, abscesses, tumors, fractures, arthritis, and "traumatic" tetanus). Syphilis and other urinary infections constituted 50% of the six major diseases affecting the *indigènes* and 30% of the six major diseases affecting the Europeans between 1904 and 1910. Each year sexually transmitted diseases seriously affected the health of hundreds of men and women in the port-city.

Table 1.2. Epidemics and Diverse Scourges, Senegal 1859–1927

| DATE | DESCRIPTION | DATE | DESCRIPTION |
| --- | --- | --- | --- |
| s.d. | Malaria | 1888–1889 | Smallpox; Famine |
| 1859 | Yellow fever | 1889–1890 | Dengue |
| 1864 | Famine; Smallpox | 1890–1891 | Influenza; Locust |
| 1865 | Locust | 1893 | Cholera |
| 1866 | Yellow fever; Locust | 1895 | Smallpox |
| 1867–1868 | Yellow fever; fires | 1896 | Insects [Punaises de bois] |
| 1868–1869 | Cholera | 1897–1898 | Drought |
| 1870 | Epizootics | 1900 | Yellow fever |
| 1872 | Drought; Famine; Epizootics | 1912 | Inundation |
| 1878 | Yellow fever | 1914 | Bubonic plague; World War I; Crisis of subsistence |
| 1881–1882 | Yellow fever | 1917–20 | Bubonic plague |
| 1882 | Famine | 1927 | Yellow fever |
| 1886–1888 | Fléau des chiques | | |

Sources: Archives of the Congrégation du Saint-Esprit (ACSE), *Bulletin Général*, 1859–1898; Myron Echenberg, *Black Death, White Medicine*; Kalala Ngalamulume, "Keeping the City Totally Clean: Yellow Fever and the Politics of Prevention in Colonial Saint-Louis-du-Senegal, 1850–1914," *Journal of African History*, 45.2 (2004), 183–202. Kalala Ngalamulume, "Le Péril Vénérien: L'État Colonial Français et la Sexualité à Saint-Louis-du-Sénégal, 1850–1920," in Jean-Paul Bado (ed.), *Conquêtes Médicales: Histoire de la Médecine Moderne et des Maladies en Afrique* (Paris: Karthala, 2006), 102. a. ANS/P165/85, Minutes of the Commission for disaster victims meeting, 16 Febr. 1912.

Besides the environmental and social conditions and international commerce, unequal access to quality health care affected disease production and mortality in Saint-Louis as well. Indeed, since the main preoccupation of the colonial authorities was to protect or restore the health of the colonial troops and the elite in the military hospital, the majority of city residents had only a limited access to a second-tier health care in the *Hospice Civil* (later Civilian Hospital) initially designed for the sick poor, the elderly, indigent sick, mental patients, and prisoners, but that also provided health care for the prostitutes who were in contact with the troops, private patients who could pay for their own care, and some people whose health care charges were paid for by the municipality. Even the privileged patients admitted to the colonial hospital still faced overcrowding and the possibility of becoming victims of hospital infection. Reports from the 1890s indicate that the Military Hospital had no separate wards for patients with contagious diseases, which contributed to the spread of diseases. At the Civilian Hospital there was a direct relationship between critical shortages of nurses, overcrowding, and mortality.

This evidence underscores class and ethnic dimensions to the pattern of diseases in Saint-Louis. The poor living conditions of the working class and the underclass in terms of housing, diet, hygiene, clean water, and waste disposal help to explain high mortality rates from infectious and parasitic diseases, and to highlight the inadequacies of public health policies. Residents from the poor urban quarters with substandard water supplies and inadequate sewage disposal paid a heavier tribute to cholera epidemics than the elite group living on the city-island; the levels of smallpox vaccination and re-vaccination their children received were inadequate, and they were the last to benefit from the urban services (sanitation, piped water, sewers, garbage collection) and surveillance programs (isolation of the sick, disinfection, vaccination, cases reporting, and public education). But yellow fever and dengue fever did affect all the segments of the urban population, rich and poor, young and adult, as the viruses and vectors carrying them found favorable conditions for their spread in the urban growth and the problems it created. Concerning the racial and ethnic dimensions to disease pattern, the eyewitness accounts indicate that the Europeans suffered more from heat and humidity than the Africans during the summer months, that the latter did not tolerate the dry and cold weather between January and May, and that the visiting Moors from the Mauritanian zone found humidity unbearable.[69]

Substantial reduction in mortality took place after 1900 due to effective solutions to the city's health problems in the areas of vector control, water supply, and sewage disposal. These changes corroborate the thesis defended by European and American urban historians that it was the improvement in the urban environment, rather than medicine or the organization of the health care delivery, that contributed to low mortality rates.[70]

## Hospitals

Being a central place for governance and services in French West Africa, Saint-Louis had the best medical infrastructure in the regions controlled by the French in West and Central Africa, if one considers the standards of the time. The creation of the Western health services came later in French Soudan (1892) and Dahomey (1893). It was the colonial medical personnel from Senegal that run such places as Guinée, including the Southern Rivers, and the French posts in Côte d'Ivoire. Even in this case, there was no regular colonial medical service outside Grand Bassam and Conakry.[71]

The history of Saint-Louis' medical infrastructure can be divided into three periods: an early period (1659–1819); a military period (1819–1897); and a public health period (1897 on). The early period started in 1659 with the building of the Fort of St.-Louis, and lasted until approximately 1819. During this period, there was no separate building for the care of the sick. One room was reserved for the sick. But with the passing of time, a separate building was made available to accommodate an increasing number of patients.

The first aides to the physicians and surgeons in the hospital were seven Sisters, members of the *Soeurs de Saint-Joseph de Cluny*, who arrived in Senegal in 1819. Their religious order was founded in 1812 by Sister A.-M. Javouhey. The Sisters' association with the Navy hospitals went back to an agreement made in 1815 between Mother Javouhey and the Minister of Interior, Lainé, allowing the Sisters to become members of the Navy hospitals labor force in some of the French overseas colonies (Reunion, Senegal, Martinique, Guyana).[72] The Sisters were answerable to the military hospital administrators for the following areas: internal economy of the hospital; custody and maintenance of linen and furniture; cleanliness of rooms; food preparation and distribution; supervision of the cooks, nurses, and other agents.[73]

The second period, which can be called a "military period," lasted until 1897. The "military" character of this period relates to the fact that the main preoccupation of the policymakers was to protect or restore the health of the colonial troops, both European and African, although the latter were less protected than the first because of their presumed acquired immunities against "tropical diseases." The Navy military health officers (surgeons and physicians) also provided health care to the administration employees and the businesses. The military period saw the development of the nursing care and the enlargement of the hospital. The conquest of most of what became known as French West Africa took place during this period. By 1897, the Sisters supervised a total of 20 nurses and 18 hospital workers (15 janitors, 2 cooks, 1 aide-cook). Overcrowding remained the main problem facing medical institutions for many decades to come.[74]

While the colonial government concerned itself with the provision of health care to the troops and the elite, the city authorities focused on the welfare of the poor, whose vote they needed to stay in power. They pushed for the creation of a medical facility for the poor. The development of the *"Annexe de Saint-Louis,"* a form of medical center, into a *Hospice Civil* and, later on, into a *Hôpital Civil* (Civilian Hospital) must be understood in this context.[75]

The public health period started in 1897, when the emphasis began to shift from military concerns, as the conquest of West Africa was over, to the general welfare of the population at large. The decree of 10 March, 1897, related to the administration of the hospitals in the colonies, constituted an important landmark opening a new era. Indeed, the decree transformed the Military Hospital into a Colonial Hospital, and opened its doors to all the categories of patients (employees from local and municipal services, businesses, and self-employed individuals and their families' members, and ordinary people who could afford it). It also reduced the cost per in-patient and out-patient day by a third.[76] The process of transformation of the military hospital into a colonial hospital took eight years to complete and it became fully effective only in 1905.

The public health period also witnessed the creation of the *Assistance Médicale Indigène* (1905), which allowed the indigent access to free consultation in all the dispensaries and, in some instances, free dressing bandages, purgative drugs, and drugs against syphilis and malaria.[77] This period also saw the creation of the Hygiene Service in 1905 to improve the urban environment and to execute the great programs of sanitation (1905–1913) discussed in chapter five.[78]

The patients admitted in the Colonial Hospital were divided into three categories: the first and second categories (Europeans, the "assimilated," and some privileged indigenous inhabitants) were admitted in the rooms located upstairs, while the rest of the *indigènes* were housed on the first floor.[79] The cost per day per case in 1897 was 12.83 francs for the officers and 8.56 francs for the non-officers.[80] The Colonial Hospital had the capacity of 283 beds, including three to four beds reserved to women in *la salle des dames*. There was no room for children.[81]

None the less, despite all its deficiencies, Saint-Louis' main hospital from all accounts was a showpiece. All the dignitaries visiting the city (bishops, African chiefs from the interior and beyond, officers, and politicians) were invited to make a tour of the hospital.

## Medical Personnel

The hospital force was dominated by Navy surgeons, or *chirurgiens de marine*, until 1867, and physicians, or *médecins*, thereafter. Navy surgeons were actually health officers trained in Navy schools of medicine at Brest, Cherbourg, and

## Table 1.3. The Medical Staff in Senegal, 1857–1867

| YEAR | SURGEON 1ST CLASS | SURGEON 2ND CLASS | SURGEON 3RD CLASS | AUXILIARY SURGEON | TOTAL |
| --- | --- | --- | --- | --- | --- |
| 1857 | 2 | 14 | 6 | – | 22 |
| 1858 | 2 | 17 | 6 | – | 25 |
| 1859 | 2 | 19 | 5 | 2 | 28 |
| 1860 | 2 | 18 | 4 | 3 | 27 |
| 1861 | 2 | 18 | 4 | 2 | 26 |
| 1862 | 3 | 17 | 4 | 2 | 26 |
| 1863 | 3 | 19 | 4 | 2 | 28 |
| 1864 | 3 | 18 | 4 | 1 | 26 |
| 1865 | 4 | 17 | 4 | 1 | 26 |
| 1866 | 4 | 16 | 4 | 1 | 25 |
| 1867 | 4 | 15 | 3 | 2 | 24 |

Sources: *Annuaire de la Marine et des Colonies* (Paris: Imprimerie Impériale), 1857, 466–67; 1858, 467; 1859, 452–53; 1860, 456–57; 1861, 499; 1862, 537; 1863, 3; 1864, 589; 1865, 608–9; 1866, 639–40; 1867, 702–3.

Toulon, and who had acquired their experience by working aboard Navy ships in the ports or colonies. They did not have the knowledge of physiology and biochemistry that doctors had. They were actually health technicians. The available data cover the period between 1857 and 1894, when the structure of the medical personnel underwent a significant change. Between 1894 and 1920 there was no radical change in this structure.

There is no evidence concerning the years between 1868 and 1872. When the data became available in 1873, a new system was already in place. The only medical staff authorized to provide health care were the *médecins de marine* (Navy physicians); *chirurgiens* who wanted to become physicians had to attend civilian *facultés* of medicine in Paris, Montpellier or Bordeaux and write a doctoral thesis.[82]

An important observation needs to be made here. The data presented in this table suggests that "Senegal" had sufficient medical staff. But, in reality, Senegal supplied other French posts with medical personnel: French Soudan and Gabon until 1891, French Congo until 1892, Dahomey until 1893; while other places such as French Guinée and Dependencies (Ivory Coast, Southern Rivers) continued to rely on Senegal until a later date. One of the consequences of this state of affairs was that the lack of sufficient physicians and surgeons became a permanent prob-

## Table 1.4. The Medical Staff in Senegal, 1873–1894

| YEAR | MÉDECIN PRINCIPAL | MÉDECIN 1ERE CL | MÉDECIN 2E CL | MÉDECIN AUXIL. | TOTAL |
|---|---|---|---|---|---|
| 1873 | 1 | 4 | 3 | 14 | 22 |
| 1874–1876 | 1 | 4 | 4 | 12 | 21 |
| 1877 | 1 | 4 | 8 | – | 13 |
| 1878 | 1 | 4 | 3 | 11 | 19 |
| 1879 | 1 | 4 | 3 | 9 | 17 |
| 1880–1882 | 1 | 4 | 3 | 11 | 19 |
| 1883–1885 | 1 | 4 | 7 | 13 | 25 |
| 1886 | 1 | 4 | 7 | 12 | 24 |
| 1887 | 1 | 4 | 7 | 16 | 28 |
| 1888 | – | 1 | 11 | 1 | 13 |
| 1889 | – | 1 | 12 | – | 13 |
| 1890 | 1 | 6 | 34 | – | 41 |
| 1891–1892 | 1 | 5 | 26 | – | 32 |
| 1893 | 1 | 3 | 14 | – | 18 |
| 1894 | 1 | 2 | 19 | – | 32 |

Sources: *Annuaire de la Marine et des Colonies* (Paris: Imprimerie Impériale), 1873, 573; 1874, 579; 1875, 614–15; 1876, 630–31; 1877, 646–47; 1878, 651–52; 1879, 637–38; 1880, 636; 1881, 646; 1882, 663; 1883, 684; 1884, 676; 1885, 726; 1886, 751; 1887, 762; 1888, 769; 1889, 795; 1890, 834–35; 1891, 808; 1892, 806; 1893, 848; 1894, 898.

lem facing the administration. Indeed, in June 1891, Colonel Amédée Dodds, a *métis* native of Saint-Louis and *Commandant Supérieur des Troupes*, complained to the governor that Dr. Lecoeur, the only *médecin des Troupes* present in the city, had the responsibility of providing health care for patients (troops) from all the garrisons in the city as well as from camps in N'Diambor, Pointe aux Chameaux, N'Diago, and, even in the posts as far away as M'Pal and Louga.[83] The medical staff also included three to five pharmacists, who had obtained a *diplôme de pharmacien* issued by the French government, after passing the exams at the *Facultés* or *Ecoles d'Etat*.[84] Pharmacists produced and sold mineral waters, lemonades and ice.[85] In any case, pharmacists did not have the monopoly of selling drugs. They were in competition with merchants, who continued to sell some types of drugs that were believed "not to harm the public health."[86]

Table 1.5. Distribution of Catholic Nuns in Medical Facilities in Saint-Louis

| DESCRIPTION | 1897 | 1900 | 1901 | 1903 |
|---|---|---|---|---|
| Colonial Hospital | 17 | 16 | 11 | 13 |
| Civilian Hospital | 4 | 3 | 4 | 4 |
| Dispensary Ndar-Toute | 3 | 2 | 2 | 4 |
| Dispensary Sor | 1 | 1 | 1 | 2 |
| Total | 25 | 22 | 18 | 23 |

Sources: ACSE, District de Sénégambie. Etat du personnel, 1897, 1900, 1901, 1903.

When the medical infrastructure reached its full development—between 1897 and 1903—the distribution of the Catholic nuns was as shown above.

Up to 1889, the nursing staff operated without legal guidelines in the sense that there was no specific legislation organizing the nursing profession. As mentioned above, the Catholic nuns were employed as nurses, and there were also other French nurses and their Senegalese aides.[87] The nursing staff in Saint-Louis was not sufficient and, during the rainy season, the nuns worked under tremendous pressure. Some of them had to be repatriated for medical reasons; anemia and excess of fatigue were mentioned as the main causal factors.[88] In Senegal, the nuns performed their duties in both the colonial and civilian hospitals until 1904, when a decision about the separation between the State and the Church was implemented in Senegal.[89] However, they did not leave Senegal en masse. Between 1904 and 1913, ten nuns continued to run the orphans' home in Ndar-Toute and the city dispensaries in Ndar-Toute and Sor, and they continued to visit the sick in their homes.

The data from 1889 indicates nursing profession comprised two categories.[90] The nurses in the first category, including *infirmiers chefs* 1st class and 2nd class, and *infirmiers-majors* 1st class, were hired by the ministry in Paris. To be eligible, the candidates needed to be French citizens or naturalized French (males or females) younger than forty years old. The governor of Senegal hired the nurses in the second category—*infirmiers ordinaires* of 1st class and 2nd class. By making provisions for the creation of the second category of "colonial nurses," the decree opened the nursing profession to the indigenous people. A distinction was made between those with enough professional competence (*infirmiers de salle*), and the maintenance nurses less qualified (*infirmiers illetrés*).[91] In 1896 there were 19 nurses in Saint-Louis.[92]

There were reports that the French colonial nurses were of mediocre quality because of either insufficient training or their conduct. The Senegalese nurses were also reputed to be mediocre auxiliaries upon whom the hospital could not rely for an intelligent care for the sick. However, their services were appreciated during the rainy season, when the hospital was overcrowded and there was not enough nursing staff. The issue of high rates of bed occupancy was aggravated when the hospital used some rooms to house French nurses, including their family members, in the hospital.[93]

The lack of adequate medical and nursing staff, and resources were not the only problems facing the administration. The difficulty of establishing harmonious physician/patient relationship very much concerned health officers and threatened the future of the Civilizing Mission. The solution to the problem consisted of recruiting more Africans to be used as interpreters between physicians and patients.

## Dispensaries and Other Health Structures

In November 1876, on the initiative of Bishop Duboin and with the help of Mrs. Brière de l'Isle, the governor's wife, the Catholic nuns opened the Sainte Anne dispensary in Ndar-Toute in order to provide health care to the urban poor. The rent for the house used as a dispensary was paid by the administration. The dispensary was run by two nun nurses and supervised by the municipal physician. It was subsidized by the colonial administration.

Patients also came from the city-island as well as from out of town; an average of 90 patients received health care daily. The most effective cure related to eye infections and wounds. Another dispensary, called the *Annexe Notre-Dame-de-Lourdes*, was built in Sor in 1888 in the vicinity of the Church Notre-Dame-de-Lourdes in order to provide health care to the 1,500 inhabitants living in this slum; the Municipal Council provided subsidies for the dispensary. It was run by one nun and was supervised by one *médecin major* (2nd class) who was also in charge of the village for sleeping sickness patients as well as of the center for vaccine.[94] The dispensary was subsidized by the Municipal Council.[95] The need to control the spreading of sleeping sickness led the colonial administration to the decision to create a village for patients with sleeping sickness (*Village de Segrégation*) in Bouetville in Sor. By 1910, the village housed 44 patients.[96] Two years later, the administration created a center for the production of vaccine to be attached to the village.[97] But the village does not seem to have fulfilled its mission because by 1916 there was a discussion about the possibility of suppressing it due to the lack of funds.[98] Another village was established in Sor for patients with leprosy. It could accommodate 24 patients. The surplus patients were housed in the village for patients with sleeping sickness.[99]

Another element of the medical infrastructure was the prison infirmary that opened in 1866. It housed all categories of people, Europeans and Africans, soldiers and civilians, who needed to be cared for while incarcerated. The infirmary, equipped with ten beds, was located in a special, well-ventilated room. A physician and a nurse visited it twice a week.[100]

The authorities created new structures outside the city in order deal with epidemic crises: camps for the isolation of troops and lazarettos for the suspects. The main isolation camps were the following: Makana, Bop-Diarra, Richard-Toll, Pointe-aux-Chameaux, N'Diago, Dakar-Bango, and M'Pal.[101] The Navy fleet and the commercial fleet were also used as temporary hospitals during the epidemic crisis.[102]

Until 1890, mental illness was not seen as a serious medical problem; both the Military and Civilian Hospitals did not have in-patient psychiatric services. With the making of mental illness a medical problem, mental patients who were seriously ill or very disruptive were kept in a few rooms in the civilian hospital before being either transferred to France or returned to their families.[103]

The restoration of the health of the administration employees and soldiers was another area of concern for the administration. Starting in 1897, the European colonial personnel who were "debilitated" by the *hivernage*, 'colonial diseases of one sort or another,' or 'living conditions in the tropics' were offered an opportunity to take medical leave, to receive 'special treatment' in the form of 'hydro-mineral cure,' and to recover for four months in one of three thermal stations in France: Aix-les-Bains, Chatel-Guyon, and Bourboule.[104] Soldiers, Junior Officers, and the 'Assimilated' from the Army and Navy were admitted in two sessions between 1 May and 30 June and between 1 September and 31 October, while Senior Officers or the 'Assimilated' from the rank of captain up were admitted for one uninterrupted session between 1 May and 31 October.[105] In 1909 the thermal station of Dax (Landes) was added to the list.[106] How many civil servants from Senegal visited these thermal stations is not known.


The growth of Saint-Louis from a fishing village into a trading post and a city-port/capital in the framework of international commerce and travel provides a privileged terrain for understanding the social basis of health and ill health, the root cause of insalubrities, as well as the creation and genealogy of negative images of Senegal.

When in 1848 Senegal became officially a French colony and Saint-Louis its capital, the colonial administration faced major organizational problems. City growth led to population growth and congestion. The city developed without a master plan and became 'the wrong kind of city' for the Europeans because of the coexistence between the Europeans and the Africans, and the rich and the poor, the daily encounters in the streets with the 'dangerous classes,' and the 'noise.' In the second half of the nineteenth century, the colonial authorities began to use different strategies to eliminate the latter from the city center, including the building codes and the outbreaks of epidemic diseases, thus de facto excluding them from many of the benefits of the colonial city life that would be introduced thanks to the transfer of new technologies.

There was no health department to deal with serious health challenges and the physician in charge of health bureau was accountable to the Interior Director, not to the governor. It took repeated threats of the epidemics to the colonial economy, administration, and social order to see the creation of the Health Service and the improvement of the social status and ranking of the Chief Medical Officer within the colonial bureaucracy. Access to fresh drinking water during the dry season between January and May, especially for the urban poor, was problematic and a source of many diseases of the digestive system. The inability of the city government to collect garbage in the slum areas and the lack of sewerage forced the urban poor to rely on the Senegal River water that was soon transformed into a repository of waste matters, including human waste, which led to environmental contamination and river pollution. The existence of swamps, bogs, moats, and other sites of standing water provided the breeding ground for vectors of infectious diseases.

The outbreaks of epidemic diseases underlined the inadequacies of public health policies and the inefficiency of municipal services. Numerous health ordinances were never fully implemented and only a few violators were actually persecuted. The miasma theory of disease causation inspired isolation and quarantine measures that were inefficient, as well as a racialized policy of residential segregation. The building of the medical infrastructure did not result from a colonial project with clearly defined goals. Instead, it was put in place over a long period of time, following the needs of the troops, the elite, and the population at large (the ordinary people, the prostitutes, the prisoners, the elderly, the mental patients). But the problems were enormous: crowding and shortage of beds, lack of sufficient and interested medical and paramedical staff, and scarcity of resources, absence of facilities for child birth and for children, poor management, and the ignorance of the physicians concerning the causes of many diseases.

Western medical infrastructure in Saint-Louis was insufficient to accommodate the needs of the urban population in terms of routine care; there were too many patients for few rooms, and various infections could spread easily through hospital wards. The medical and nursing staff was insufficient. Thus, when the epidemics struck the city, the administration could not rely on the existing infrastructure to care for the sick.

It should be argued here that medical interventions and advances and the expansion of medical services were not the cause of the decline of mortality for different segments of population of Saint-Louis, as the apologists of the "Civilizing

Mission" claimed it. Instead, the most important factors at work in the city were the sanitation and improvement of the urban environment, the provision of clean drinking water and safe food, improved housing and sewage disposal, and change in individual behavior. For as long as patients who received treatments went back to live in the same damaged environment and to drink the same dirty water, no health improvement could be expected.

But beyond the medical practice are the discursive practices and representations of the Other, the circulation of images of Senegal resulting from Saint-Louis' experience with insalubrious conditions, unbearable stench here and there, and epidemic diseases. The intersection of the environment, race, disease, ethnicity, and religion (Islam and African Traditional Religions, Islam, and Christianity) shaped the medical, colonial and missionary discursive practices that "tropicalized" Senegal, that is, constructed it as a land of heat and disease and a land cursed by God, and stigmatized and pathologized the local population. The following chapter examines the ways in which the intersection of the environment, race, and yellow fever epidemics shaped racialized health policies and racially oriented colonial and missionary discourses.

## Notes

1. For the relationship between Saint-Louis and its hinterland, see Boubacar Barry, *Le Royaume du Waalo: Le Sénégal avant la conquête* (Paris: François Maspero, 1972).
2. Anne Perotin-Dumon, "Cabotage, Contraband, and Corsairs: The Port Cities of Guadeloupe and Their Inhabitants, 1650–1800." in Franklin W. Knight and Peggy K. Liss, eds., *Atlantic Port Cities: Economy, Culture, and Society in the Atlantic World, 1650–1850* (Knoxville: University of Tennessee Press, 1991), 62; A. W. Lawrence, *Trade Castles and Forts of West Africa* (London: Jonathan Cape, 1963), 78.
3. ANFOM/SGEOS/XI/27, Dr. Adolphe Salva, "Appendice médicale concernant les maladies de la colonie du Sénégal et de ses Dépéndances.
4. Jeffrey D. Needell, "The Revolta Contra Vacina of 1904: The Revolt against 'Modernization' in Belle-Époque Rio de Janeiro," *The Hispanic American Historical Review*, Vol. 67, No. 2 (May, 1987), 233–269.
5. Andrew L. Knaut, "Yellow Fever and the Late Colonial Public Health Response in the Port of Veracruz," *The Hispanic American Historical Review* 77, 4 (Nov., 1997), 619–644.
6. Georges Courreges and Fadel Dia, *Saint-Louis du Sénégal* (Clermont-Ferrand: Editions Soprep, 1982).
7. Archives of the Congrégation du Saint-Esprit, *Journal de la Communaute de Saint-Louis, 1904–1951*, entry for May 2, 1905.
8. It included a certain number of villages: Leybar, Ngalel, Maka Toubé, Gandon, Ngay-Ngay, Ndiaker, Mouit, Ndiol, and Ndieben. See Y.-J. Saint-Vincent, *Le Sénégal sous le Second Empire. Naissance d'un empire (1850–1871)* (Paris: Karthala, 1985), 415.

9. R. Pasquier, "Un aspect de l'histoire des villes du Sénégal: les problèmes de ravitaillement au XIXe siècle," *Cahiers du C.R.A.*, no. 5 (1987), 192; see also Yves-Jean Saint-Martin, *Les Sénégal sous le Second Empire. Naissance d'un empire colonial, 1850-1871* (Paris: Karthala, 1989), 108.
10. Most of the mid-nineteenth-century travelers observed that some of the traders from Saint-Louis used to spend six or seven months per year doing commerce outside the city. See, for example, L. Faidherbe, *Le Sénégal. La France dans l'Afrique Occidentale* (Paris: Librairie Hachette, 1889), p. 66; P. Soleillet, *Voyage à Segou, 1878-1879* (Paris: Challamelaîné, 1887).
11. The initial *escales* were Fort Saint-Joseph in the Galam (1700-1732), Senoudebou (established in 1715), Bakel (1818), Dagana (1819), and Richard Toll. D. T. Niane and J. Suret-Canale, *Histoire de l'Afrique Occidentale* (Paris: Présence Africaine, 1961), p. 109. The last three trading stations were established as part of a large project for "settler agriculture" conceived by the Minister of Navy Baron Portal after a period of shifting control over Saint-Louis between the British (1758-1778, 1809-1817) and the French (1778-1809). The project aimed at developing local resources with the collaboration of the local people under the supervision of Governor Schumaltz: Podor (1854), Medine (1855), Bakel (1855), Matam (1857), Ndangan (1858), Kenieba (1858, 1861), Salde (1859), Aere (1866), Kayes (1879).
12. The concessionary companies included the following: the Compagnie du Cap Vert et du Senegal (1658-1664); Compagnie des Indes Occidentales (1664-1673); Compagnie du Senegal et d'Afrique (1673-1682); Compagnie Royale du Senegal et Cote d'Afrique (1696-1709); Compagnie du Senegal (1709-1718); Compagnie des Indes (1718-1767); Compagnie d'Afrique (1774-1776); Compagnie de la Guyane (1776-1783); Compagnie Nouvelle du Senegal et Dependances (1783-1791). The companies received commercial monopoly in exchange for administrative and military expenses incurred until 1763; from then on, a royal governor took over administrative and military powers. Companies were accountable to the Navy Secretary in charge of colonies, the Finances General Comptroller, the Foreign Affairs Ministry, and the shareholders in Paris, Rouen, Nantes, La Rochelle, and Bordeaux. The monopoly was suppressed in 1791 by the French Constituante and restored in 1824 by the merchants from Saint-Louis, who created the Compagnie de Galam in 1824. The company had the monopoly of trade in the Upper Senegal River between January and August. For more details see Regine Bonnardel, *Saint-Louis du Senegal: Mort ou naissance?* (Paris: L'Harmattan, 1992), 30-31; E. Saulnier, *La Compagnie de Galam au Sénégal* (Paris: Emile Larose, 1921).
13. Leland Conley Barrows, "General Faidherbe, the Maurel and Prom Company, and French Expansion in Senegal." Ph.D. thesis, U. of California, Los Angeles, 1974, 50.
14. It included the bureaus of the Ordonnateur, Administrative Services (1840), Colonial Inspector; the Directions of External Affairs (1846-1863, then Political Affairs, thereafter)—organized like the Algerian Bureaux Arabes—and of Interior (1869-1872, 1882-1898, then General Secretariat thereafter) resembling the Guyanese model.
15. See G. E. Brooks Jr., "The *Signares* of Saint-Louis and Gorée: Women entrepreneurs in Eighteenth-Century Senegal," in N. J. Hafkin and E. G. Bay (eds.), *Women in Africa. Studies in Social and Economic Change* (Stanford, CA: Sanford U. Press, 1976), 19-44.
16. Francois Zuccarelli, *La vie politique sénégalaise (1789-1940)* (Paris: Cheam, 1987), 12.
17. ANS, Dr. A. Corre, *Moniteur du Sénégal et Dépendances*, Febr. 23, 1864, 1, 12; Febr. 23, 1864, 22; May 24, 1864, 55. See ANS/3G3/4/306, letter of Sept. 9, 1864; ANS/3G3/4/309, decree of May 24, 1865 organizing the assistance to the poor. See also ACSE, Bulletin de la Congrégation

des Soeurs de Saint-Joseph de Cluny, IV, 303; Boîte no. 146 A, Sénégal IV: Sénégambie. Histoire du Sénégal 1845–1898.
18. ANS/AOF/22G6/39, Récensement de la population, 1877.
19. F. Brigaud and J. Vast, *Saint-Louis du Sénégal. Ville aux mille visages* (Dakar: Ed. Clairafrique, 1987), 37–38.
20. F. Zucarelli, *La vie politique sénégalaise*, op. cit., 14.
21. Fréderic Carrère and Paul Holle, *De la Sénégambie française* (Paris: Librairie de Firmin Didot Frères, 1855), 16; for more information see D. Robinson, *Paths of Accommodation: Muslim Societies and French Colonial Authorities in Senegal and Mauritania, 1880–1920* (Athens: Ohio U. Press, 2000), esp. chaps. 5–6; J. Hilary, "French Citizens or Imperial Agents? The *Métis* of Saint-Louis and Republicanism in the Colonial Period, 1871–1920," Ph.D. dissertation, Michigan State U., 2003.
22. *Echo de Saint-Louis*, no. 25, Nov. 1971, 5.
23. M. Marcson, "European-African Interaction in the Precolonial Period: Saint-Louis, Senegal, 1758-1854," Ph.D. dissertation, Princeton U., 1976, 34, 39.
24. For more details on status privileges and symbols, see Holger R. Stub (ed.), *Status Communities in Modern Society. Alternatives to Class Analysis* (Hinsdale, IL.: The Dryden Press, 1972).
25. F. Carrère and P. Holle, op. cit., 16-17.
26. The advertisements in the Journal Officiel du Sénégal et Dépendances contain the names of the landlords and the buildings to be rented. In addition, other sources, such as the Bulletin Administratif, provide insights into the conditions of accommodation in the city. For example, in 1857 Mrs. Charbounié rented her house located on 28 Cornier Street to the Spahis for 2,000 francs per year (see Bulletin Adm., 107–8). In December 1885 the Navy Commandant paid 4,500 francs per year to rent a house that was used for office as well as for residence; for details, see Bulletin Administratif, 1885, ministerial dispatch, Dec. 4, 1885, 532–34.
27. H25/AOF/24, F. de Coutouly, "Notes sur les *métis* en A.O.F.," Bulletin de la Société des Anciens élèves de L'École Coloniale, no. 79, mai 1918, 4–12.
28. Mamadou Diouf, "Traitants ou négociants? Les commerçants Saint-Louisiens (2e moitié du XIXe siècle–Début XXe siècle): Hamet Gora Diop (1846–1910). Etude de cas." Paper presented to the Colloque sur les Grands Commerçants Africains de l'Afrique Occidentale, Dakar, 1–4 May, 1990.
29. R. Pasquier, "Un aspect de l'histoire des villes du Sénégal: les problèmes de ravitaillement au XIXe siècle," *Cahiers du C.R.A.*, no. 5, 1987, 192.
30. David Robinson, *Paths of Accommodation. Muslim Societies and French Colonial Authorities in Senegal and Mauritania, 1880–1920* (Athens, Ohio U. Press, 2000), 122–130.
31. Tita Mandeleau, *Signare Anna, ou le voyage aux escales* (Dakar: NEAS, 1991), 15–16.
32. Louis Chevalier, *Classes laborieuses et classes dangereuses à Paris pendant la première moitié du XIXe Siècle* (Paris: Plon, 1958).
33. Interview with Omar Sarr, Saint-Louis, Apr. 7, 1994.
34. ACSE, Chevilly-Larue, France. "Dossier 159B, V, Sénégal, Affaires Diverses." Montel to Mgr. Riehl, Jan. 9, 1886; Bulletin de la Congrégation des Soeurs de Saint-Joseph de Cluny, 1886–7, 233; Journal de Saint-Louis du Sénégal, 1852–1904, entry for Oct. 13, 1888.
35. ANFSOM/P4, "Protection du Littoral de Saint-Louis," Rapport de M. Giraud, Febr. 1, 1951, 1–2.

36. ANS, "Arrêté portant agrandissement du village de Bouetville, July 25, 1864, Moniteur du Sénégal et Dépendances, Aug. 2, 1864, 95.
37. Yves-Jean Saint-Vincent, *Le Sénégal sous le Second Empire. Naissance d'un empire colonial (1850–1871)* (Paris: Karthala, 1985), 107.
38. ANS, decree of Dec. 30, 1884, Bulletin Administratif, 193.
39. F. Zuccarelli, *La vie politique sénégalaise (1789–1940)* (Paris: Cheam, 1987), 19.
40. Alain Coursier, *Faidherbe, 1818–1889: Du Sénégal à l'Armée du Nord* (Paris: Tallandier, 1989), 94.
41. Ministère de l'Urbanisme et du Logement, *Urbanisme et habitat en Afrique noire francophone avant 1960* (Paris: Agence Française pour l'Amenagement et le Développement à l'Etranger, April 1984), 598.
42. ANS, Moniteur du Senegal et Dependances, no. 41 of January 6, 1857.
43. ANS/3G3/4/302, Mayor Deves to governor, no. 97 of March 11, 1876.
44. ANS/3G3/4/346, mayor to Commandant Supérieur, no. 207 of May 23, 1891.
45. ANS, municipal ordinance of May 12, 1897; the ordinance was approved by the *Conseil Privé* on May 15.
46. F. Carrère and P. Holle, *De la Sénégambie Française* (Paris: Librairie de Firmin Didot Frères, 1855), 16.
47. For the political, economic, spatial and cultural importance of the built environment, see A. King, *Global Cities. Post-Imperialism and the Internationalization of London* (New York: Routledge, 1990), 11.
48. ANS/H?, pétition of "Mères de Familles de Saint-Louis" to governor, 15 March 1887.
49. *Le Réveil du Sénégal*, July 26, 1885, 3.
50. Archives Municipales, I A 63; Moniteur du Sénégal et Dépendances, 264–65.
51. G. W. Johnson, *Naissance du Sénégal contemporain. Aux origines de la vie politique moderne, 1900–1920* (Paris: Karthala, 1991), 53, 136–37.
52. Archives Municipales de Saint-Louis, 1Q80: recensement général de la population, Febr. 20, 1914.
53. Annales, 144; G. W. Johnson, *Naissance du Sénégal contemporain*, 64–65.
54. ANS, Bulletin Administratif, 1892, 195.
55. Journal Officiel du Sénégal et Dépendances, July 1, 1893, 18.
56. G. W. Johnson, *Naissance du Sénégal contemporain*, 146–47, 149, 163.
57. David Robinson, *Paths of Accommodation*, 1.
58. ANFSOM, Série Géographique Sénégal XI d. 32, Inspecteur Général du Service de Santé de la Marine, report of Oct. 13, 1831. The death of eleven soldiers was attributed to 'nostalgia,' which prompted the General Inspector to recommend 'sending only men who come there voluntarily.'
59. Pierre Loti, Le Roman d'un Spahi.
60. ANS/Senegal/2G3/18. Annual Report, 1903, 114.
61. The decree of 23 Dec. 1897 had fixed the duration of service at three years.
62. ANS, Circular of 15 Febr. 1910, Journal Officiel du Sénégal, 128.
63. ANS/AOF/H8/124, médecin en chef, Michel, to governor, no. 20 of 9 Aug. 1890; see also ACSE, "Journal," entry for 11 July 1890, 247. For general information, read World Health Report, 1996, 48–50; see also J. Lederberg, R. E. Shope, and S. C. Oaks Jr. (eds.), *Emerging Infections: Microbial Threats to Health in the United States* (Washington, DC: Institute of

Medicine, National Academy Press, 1992); Sir S. Rickard Christophers, *Aedes Aegypti (L.) The Yellow Fever Mosquito*, 57–8, 81–2.

64. Sometimes, the river water remained salty until August. See Moniteur du Sénégal, 2 Aug. 1868, 218; ANS/Senegal/H20, the Hygiene and Public Sanitation Commission meeting records, 10 Jan. 1868; see also ANFSOM/Séries géogr./Senegal/XII/17 b, "Memoire sur un avant-projet de conduite d'eau destinée à l'alimentation de Saint-Louis et de ses faubourgs," 6 Apr. 1869 by the chief of Bureau du Genie.
65. ACES, Bulletin Général, 1868, 862.
66. ANFSOM/Séries géographique/Senegal/XII/17 b.
67. ANS/AOF/H2/38, médecin en chef to governor, n.d.
68. Achives municipales de Saint-Louis 2 B 20, Acting Interior Director, L. Turquet, to mayor, no. 80 of 26 May 1888; ANS/AOF/H39, Police deputy chief, Pellissier, to Interior Director, no. 72 of 13 June 1888.
69. R. Bonnadel, *Saint-Louis du Senegal: Mort ou naissance?*, 51.
70. See, for example, Judith W. Leavitt, *The Healthiest City. Milwaukee and the Politics of Health Reform* (Princeton, NJ: Princeton U. Press, 1982); Anthony S. Wohl, *Endangered Lives. Public Health in Victorian Britain* (Cambridge, MA: Harvard U. Press, 1983); Christopher Hamlin, *Public Health and Social Justice in the Age of Chadwick, Britain, 1800–1854* (New York: Cambridge U. Press, 1998).
71. ANS/H24/Senegal, Record of the Sanitary Commission meeting, 6 June 1899; see also Annuaire Coloniale, 1891, 812; 1893, 853.
72. Anonymous, *Une grande missionnaire Anne-Marie Javouhey, 1779–1851* (Osny, 1990), 9–10.
73. Dr. P. Aude, *Code des Officiers du Corps de Santé de la Marine* (Paris: Berger-Levrault et Cie, 1877), 81–7.
74. ANS/AOF/H2/37, médecin to Ordonnateur, n.d.; see also ANS/AOF/H2/39, médecin to Governor, n.d.
75. The available evidence concerning the history of the *Hospice Civil* comes from various sources, but the most important source is a small brochure (notice) written in 1900 by Dr. Charles Carpot, a *métis* native of Saint-Louis who got his education in France. Entitled "*Dix ans de nosologie à l'Hôpital Civil de Saint-Louis*," the brochure covers a ten-year period of medical activities at the hospital (1889–1898).
76. ACSE, "Bulletin," LVI Dec. 1899, 669.
77. ANS/2G13/26, Annual report 1913, 63.
78. See "Le secrétariat général du gouvernement général. Un emprunt de 65 million," Questions diplomatiques et coloniales, XV (1903), 195-96; Arrêté no. 742, Journal Officiel de l'Afrique Occidentale Française, 1905, 509-12; Arrêté no. 152, Journal Officiel du Sénégal, 1907, 87-88; Arrêté no. 219, Journal Officiel de l'Afrique Occidentale Française, 1909, 104-7; cited by Margaret Osborne McLane, "Economic Expansionism and the Shape of Empire: French Enterprise in West Africa, 1850–1914," Ph.D. dissertation, University of Wisconsin–Madison, 1992, 328–32.
79. ANS/2G13/26, Rapport annual 1913, 62.
80. Dr. Primet, "Rapport sur l'application au Sénégal du réglement du 10 Mars 1897," 7.
81. Dr. Primet, "Rapport sur l'application au Sénégal du réglement du 10 Mars 1897," 5–6.
82. See Yves-Jean Saint-Vincent, *Le Sénégal sous le Second Empire. Naissance d'un empire colonial, 1850–1871* (Paris: Ed. Karthala, 1989), 142 note 19.

83. ANS/H8/AOF/12, Commandant Supérieur des Troupes, Colonel Dodds, to Governor, no. 689 of 28 June 1891.
84. See Arrêté of the governor general of 25 May 1906, Bulletin Administratif, 1906, 553–61.
85. ANFSOM, Série géogr. Sénégal XI d. 36, governor to ministry, 9 Oct. 1883.
86. ANFSOM, Série géogr. Sénégal XI d.36, Caminale (pharmacist) to governor, March 1875.
87. ANS/H2/94 (A.O.F.), médecin to Governor, 2 September 1867.
88. ACSE, Paris 14, Senegal 2A E8, certificats de visite médicale, 1886–9.
89. ANS, Journal Officiel du Sénégal et Dépendances, 16 Jan. 1904: Circulaire ministérielle, 28 Nov. 1903, 30.
90. ANS, Journal Officiel du Sénégal et Dépendances, supplément, 11 April 1889, 147. The decree of 11 April 1889 and the ministry dispatches dated 5 June 1893 and 20 March 1894 organized the nursing profession.
91. Dr. Primet, "Rapport sur l'application au Sénégal du réglement du 10 Mars 1897 sur le fonctionnement des hôpitaux,." 14.
92. ANS/2G1/7, Annual report 1896, 17; see also Service de Santé, Rapport sur les économies susceptibles d'être réalisées dans les dépenses du service hospitalier au Sénégal (dépêche du 23 Oct. 1896), 2.
93. ANS/2G1/7, Annual report 1896, 17–18.
94. ANS/2G13/26, Annual report Senegal, 139.
95. Archives Municipales de Saint-Louis, E 94, Municipal Council meeting, 25 July 1895.
96. ANS/2G10/26, Annual report 1910, 56.
97. ANS, Annual report 1912, 264.
98. ANS/2G16/18, Rapport médical annual 1916, by Dr. Damien, 30.
99. ANS/2G16/18, Rapport médical annual 1916, by Dr. Damien, 27, 30-1.
100. ANFSOM, S. géogr. Sénégal XI d. 21 A, governor to minister, 9 August 1884; C, governor to under-Secretary of colonies, 7 Nov. 1892
101. ANS/H52/2/AOF, governor general to ministry of colonies, no. 605, May 16, 1905.
102. ANS/H2/III/AOF, médecin to ordonnateur, Oct 9, 1867.
103. See Kalala Ngalamulume, "Classify and Sequestrate: The Regulation of Madness in Saint-Louis-Du-Senegal, 1890–1914," in Kalala Ngalamulume and Paula Viterbo (eds,), *Health and Medicine in Africa* (Berlin: LIT Verlag, forthcoming).
104. For more details read Eric T. Jennings, *Curing the Colonizers: Hydrotherapy, Climatology, and French Colonial SPAS* (Durham, NC: Duke University Press, 2006).
105. ANS/H7/Senegal, The Société des Eaux Minérales de Chatel-Guyon Administration Council President to governor of Senegal, June 6, 1905.
106. See the decree of 6 February 1909 in the Journal Officiel du Sénégal, 135.

CHAPTER TWO

# The "White Man's Disease": The Great Yellow Fever Epidemics, 1867–1900

The French experience with yellow fever in Senegal will be told in two parts. The first part is about the "war" led by French physicians, guided by the miasmatic theory of disease causation to make sense of and to explain the disease environment of tropical Africa, against an enemy that remained elusive, yet lethal; the construction of discursive practices that labeled yellow fever as the "White man's disease" and elaborated the image of Senegal as a "tropical" land and a land of heat and disease; and the adoption of specific strategies to combat yellow fever epidemics. Because of their role in leading the war against yellow fever, physicians were awarded in 1883 with the creation of an autonomous Health Board, directly accountable to the governor instead of the Interior Director, and in 1892 with the membership of the *Médecin en Chef* into the Defense Council (*Conseil de Défense*) and Private Council (*Conseil Privé*). This story is told in this chapter. The second part of the story will focus on the conflict between the interests of commerce, public health and civil liberties, on the one hand, and that between the supporters of the environmentalist paradigm and those of the new bacteriology of Pasteur and Kock promoting new approaches to finding explanations to the mystery of yellow fever and other "pestilential exotic diseases" (especially, cholera and plague), and to fighting them. But this story cannot be told here for chronological reasons since during the same period Saint-Louis also experienced two serious outbreaks of cholera epidemics, respectively in 1869 and 1893, a disease that was constructed as the "Black man's disease," and that added to the drama. All these developments made it difficult to tell the story of Saint-Louis' experience with yellow fever epidemics in one chapter. This is the reason why chapter 3 will deal with cholera epidemics and the narrative of yellow fever will continue in chapter four.

In 1867, Saint-Louis was struck by the first of great yellow fever epidemics that killed or disabled thousands, disorganized the operations of the economy and the administration, and terrorized the urban residents. The official response went beyond the sanitary and medical measures to include blame and medical profiling of the urban poor. The implementation of these measures was met with resistance not only from the urban poor but also from the merchant community and the French populace in general. Missionaries responded differently from their compatriots; they understood and explained the ravages of yellow fever in terms of punishment from God. Yellow fever left a deep imprint on the psyche of the city residents, who came to mythologize and personify the disease.

## The Ecology of Yellow Fever

Yellow fever is an endemic disease of forest monkeys in Africa and the Americas, where it circulates between monkeys and certain species of mosquitoes, especially the *Aedes Africanus*. As a result of deforestation and ecological changes, there is an occasional mosquito-to-human transmission by female *Aedes aegypti* mosquito vectors, which thrive in areas of standing water around human settlements, such as cisterns, discarded cans or tires, and containers of clean waters with solid sides to which they can cement their eggs.

The lifespan of the female *Aedes aegypti* is around two months. Infected mosquitoes can also travel in ships and trains and spread the disease. Such transmission can lead to an epidemic, or urban yellow fever. The incubation period before the onset of symptoms varies between three and seven days after infection during which time the virus "invades through the lymphatic system and rapidly proliferates in visceral tissue cells. The liver, kidneys, and heart are principally involved, and the degenerative changes in those organs are responsible for most of the clinical symptoms."[1] The symptoms are terrifying and cause general panic: jaundice, high fever, internal hemorrhage, and vomiting of black blood. Between one half and a third of infected people die, others recover. Yellow fever thrives where there is adequate rainfall and warm weather, with temperature varying between 70° and 90° F. Today, outbreaks of urban yellow fever still occur in South America between January and April and in some regions of West Africa between May and October.

It has been argued that the fact that symptoms were mild in children conferred immunity to permanent residents of endemic areas or areas frequently visited by yellow fever, thus restricting the outbreaks of epidemics to the sudden arrival of large groups of newcomers such as immigrants, civil servants, soldiers, sailors, and so forth. This view can be found in P. Curtin's assertion that "yellow fever in West Africa was a strangers' disease, attacking those who grew up elsewhere."[2] Curtin's interpretation was a reflection of official mortality statistics that provided more details about the Europeans and the "assimilated" group than about the majority of the natives, who had limited access to medical services. But recent scholarship in molecular biology has revealed that a person who has been exposed to a viral or bacterial infection acquired protective "immunity" not against re-infection, but against severe or life-threatening symptoms.[3] Two conclusions can be drawn from these observations. First, the "immunity" that the *indigènes* had acquired must be understood in relative terms. They did get sick but remained far away from European eyes for lack of access to colonial medical institutions, which helps explain their invisibility in the medical records. Second, despite the exaggerations, the fact remains that yellow fever was a lethal threat to the European newcomers who were first exposed as adults and lacked antibodies acquired in childhood.

Although contemporaries had no doubts that yellow fever originated in Central and South America, medical historians are divided concerning whether the origin of yellow fever is in Africa or the Americas. The proponents of the American origin argue that travel accounts described yellow fever epidemics there a century before it was recognized in Africa. Those who argue in favor of the African origin hypothesis, however, point to the susceptibility to disease of both the Indians and American monkeys as a refutation of a long history of exposure, while the West Africans' ability to resist yellow fever is an indication of acquired immunity.[4] In any case, international trade and travel contributed to the spread of yellow fever in Africa, the Americas, the Caribbean, and Europe.

A Cuban physician, Dr. Carlos Finlay, identified in 1881 a mosquito vector as its agent of transmission, but his theory remained in scientific limbo until it was verified in 1900 by J. Lazear; the virus itself was discovered in 1901 by Dr. Walter Reed, and both Lazear and Reed were members of the American Yellow Fever Commission in Cuba.[5] The first vaccine was developed in 1937.[6] Before these discoveries yellow fever remained the most feared disease. It had swept through the Caribbean, Brazilian coastal towns, the coast of West Africa, Louisiana and the Mississippi Valley since the eighteenth century. It struck Barcelona in 1821 where it killed 20,000 residents. Yellow fever was signaled on the coastal regions of West Africa in 1766 and 1778–1779. It ravaged Gorée Island in 1830, reached Lisbon in 1857–1858, and returned to Senegal in 1859 and 1866 without reaching the epidemic level in Saint-Louis. An epidemic is defined as "a marked rise in the frequency of a specific infectious disease in a community over a limited period."[7] From 1867 on, yellow fever became a lethal threat to the city, especially during the rainy summer months.

But in actuality, as indicated in chapter 1, it was malaria that killed more people than yellow fever, in part because, as James Webb Jr. put it,

> the mosquitoes that are the principal vectors for malaria on the African continent, *Anopheles gambiae* and *Anopheles funestus*, developed marked preference for human blood—rather than animal or bird blood—and became the most efficient of the world's malaria-transmitting, anopheline vectors.[8]

Another reason has to do with the fact that the etiology of yellow fever, as explained above, is far less complicated than that of malaria.

In nineteenth-century Senegal, yellow fever occupied a special place in the minds of French authorities and colonists because of the recurrence and disruptive nature of the epidemics, the high mortality rates, the panic and flight that it provoked, the psychological impact it made on people's minds, but also because of the general perception that it was the "White man's disease" par excellence, and the challenge that it posed to the rhetoric of the *Mission Civlisatrice*. The legitimating discourse of the colonial officials emphasized liberating the natives from the tyranny of disease over health, of instinct over reason, and of ignorance and superstition over knowledge, without engaging what W. E. Connolly called the "enigma of otherness."[9] This "human hecatomb" contributed to the reputation of Senegal as the most dangerous colony after Martinique and Guadeloupe at the time when doctors knew very little about its etiology, transmission, diagnosis, and treatment. But mortality statistics provided more details about the Europeans and the *métis* than about the *indigènes*.[10]

## Yellow Fever and the Miasmatic Theory

In his insightful work on miasma, Alain Corbin argued that in eighteenth-century France, the formulation of "knowledge" about the etiology of diseases, in general, and "fevers," in particular, focused on the soil, putrefied matters and miasma, climate changes, and the role of the air in illness.[11] Public health experts had identified the urban refuse, sewers, and cemeteries like Saint-Innocents as potential sources of ill health.[12] Haguenot's *Mémoire sur les dangers des inhumations*, published in 1771, and Vicq d'Azyr's *Essai sur les lieux et les dangers des sépultures*, in 1775, warned about the toxic and lethal nature of the "stinking odor" or "dangerous emanations" from decomposing bodies in the cemeteries and about the danger of burials inside the churches and the cities.

French physicians in Senegal, preoccupied with the protection of the health of Europeans, tried to understand the reality of everyday life, especially the causes of ill health, through the basic terms and principles formulated in France, either by applying without modification the common medical wisdom or by adapting it to the local, "tropical" conditions.[13] They all underlined the perceived dangers of the tropical disease environments, the differential susceptibility of Whites, Blacks and *métis* to different diseases, and the possibility of "acclimatization" through limited "seasoning," appropriate lifestyles, diet, hygiene, and medicine. Physician L.-J.-B. Bérenger-Féraud devoted a great deal of time to the study of yellow fever. While Chief Medical Officer in Saint-Louis from May 1871 through November 1873, he gathered the data for his prize-winning book, entitled *Epidemies de fièvre jaune au Sénégal* in 1872, and for a series of articles published in the *Moniteur du Sénégal et Dépendances* in 1873, in which he focused on the challenge posed by yellow fever in Senegal, sanitary conditions, and other weather-related ailments in Saint-Louis. He observed that, during *hivernage*, the combination of heat, thirst, fatigue, bad dreams, and mosquitoes seriously compromised the health of city dwellers. He gave a long list of hot weather–related ailments: "fevers," hepatitis, skin diseases, boils, respiratory complaints, diarrhea and dysentery, rheumatism, and eye infections (for those who spent their nights outside in the open. After he left Senegal, he published *De la fièvre jaune au Sénégal* in 1874 in which he established a link between the topography of Senegal and the emergence of yellow fever as well as other fevers. A year later, he published the first volume of his *Traité clinique des maladies des Européans au Sénégal* in which he underlined the "insalubrities of the climate of Senegambia" and its effects on the mortality of the Europeans who went to live there temporarily. He constructed Senegambia as being "different" from other regions of the globe and contended that this difference could be found in the particularities of its climate—especially its location in the intertropical zone, where it was "exposed twice a year to a perpendicular action of the sun"—characterized by two different seasons: first, *hivernage*, or the hot and humid season, and second the dry and cool season. Another special feature of the location of Senegambia had to do with the negative influences it received from the Sahara Desert in the North and East and the Atlantic Ocean in the West and from its salty clay soils.[14] He described Senegal as "one of the sorry lands, where life is a perpetual and often ineffective struggle against the disease." Thus, by the 1870s, Bérenger-Féraud and other members of the medical establishment had provided the scientific basis for the bad reputation of the "harmful climate" of Senegal.

Another medical expert, physician A. Borius, also contributed to knowledge about the effects of climate on the health of the Europeans. In *Climat et état sanitaire suivant les saisons* (1875), *Recherches sur le climat du Sénégal* (1875), and *Nouvelles recherches sur le climat du Sénégal* (1879), Dr. Borius argued that heat was "the primary cause of all other phenomena" and that there was a strong correlation between the changing seasons and diseases based on his personal observations in

Senegal; by 1882, in *Les maladies du Sénégal*, he presented a classification of diseases he believed were particular to Senegal.

Later, in 1891, Bérenger-Féraud published his *Traité théorique et clinique de la fièvre jaune*, in which he summarized the state of the "imperfect knowledge" about yellow fever in 1890, when scientists had not yet proved in any scientifically convincing way the validity of the mosquito theory of transmission of yellow fever. He became convinced that the most vulnerable among the Europeans were not the recent immigrants or those who had spent many years in the colony and were, in a certain way, already "seasoned" or "creolized," but those who were in-between, experiencing their second or third *hivernage*.[15] Thus the outbreaks of yellow fever in Senegal provided physicians, who chaired an increasing number of medical boards, with an opportunity to present a multi-causal explanation of its emergence, re-emergence and spread, including primary and secondary causes.

The epidemiological theory that provided the framework within which health authorities understood the causes of the disease and its spread included the local environmental factors, both natural and social, contagion, and individual predisposition. The determination was made on the basis of observation and experience, especially the fact that cases of yellow fever in West Africa were reported during the rainy season. The localists—variously referred to as infectionists, hygienists, sanitarians, or anticontagionists—on the one hand, argued that the causes of yellow fever had to do with local environmental conditions. They identified two elements of the natural environment as the main causes of diseases: the stagnant water and the hot or rainy season (*hivernage*, from June to November). First, French medical doctors considered the stagnant water, everywhere in the surroundings of Saint-Louis in the form of swamps, bogs, and moat, to be the source of the highly poisonous vapors called "pernicious emanations" or "miasmas" that caused all types of diseases. The northern part of the island was covered all year long by a huge swamp and the vapors rising from it were believed to infect the air. In addition, the consumption of such unclean river water was also associated with a large number of diseases, including various types of "fevers." Second, the rainy season (*hivernage*), characterized by constant variations in temperature, was directly linked to high rates of infectious and parasitic diseases, and infant mortality and morbidity. One eyewitness referred to the rainy season as the "season of the fleas" during which it is "impossible to sleep during the night."[16] The association of the rainy season with winter in France (*hiver*) was justified by the fact that it was the most difficult and depressing season for the Europeans because of the rain. Floods and stagnant water, following each heavy rain, accelerated the decomposition of all types of organic matters, which were quickly covered with abundant vegetation secreting the deadly "miasmas." For example, according to the health officials, during the 1867 yellow fever epidemic, the climate and soil, the torrential rains that had transformed the streets of Saint-Louis into infected ponds, and the landscape around the city—a flat swampy land, the mouth of the Senegal river, and the mixture of salty and non-salty water in the river around the city—provided a fertile ground for its spread and incidence.[17]

There were also two secondary causes: the pestilential emanations (miasmas) from the cemeteries and other petrified matters in soil or stagnant water, and the frequent and terrible 'winds from the south' (*harmattan*) during the hot and rainy season months. The Catholic Cemetery located in Sor at the entrance of *Pont-Faidherbe* and the Muslim cemetery located south of Guet-Ndar constituted an important source of "miasmas" and a permanent danger to the city's health, given that they were situated less than 35 to 40 meters from houses. Indeed, according to this view, when the tombs cracked under the pressure of rain—because of their low depth (0m80) and inappropriate cover—they liberated the pestilent emanations that could infect and kill those who inhaled them. These emanations were associated with the outbreak and spreading of the 1867 yellow fever epidemic.[18] And the "evil" time, when the prevailing "evil" wind was in motion and the chance for contamination high, was the evening.[19] It was believed that cemeteries remained a health hazard for seven years after their closure. Thus, the hot season, the "evil" wind, and the stagnant water on this low and swampy land at the mouth of a big river on which the city was built, constituted a dangerous trio fatal to Europeans, not to Africans or *indigènes* who were not considered to be susceptible to yellow fever.[20]

Other predisposing factors that allowed the disease to flourish included unsanitary housing, the lack of sewers and cesspits, the 'harmful' by-products of the *indigènes*' productive activities, such as waste and pollution from fish smoking and animal slaughtering, and individual psychology. The unsanitary conditions of the overcrowded dwellings shared with domestic animals were identified as "one imminent and fatal cause" of the reemergence of yellow fever in 1881.[21] In addition, rubbish, filth, urine and manure from sheep, cows, and horses, owned by the *indigènes*, in the streets and public spaces, and along the docks made an unbearable stink, which polluted the air. So localists did not see yellow fever as a disease that could be transmitted from person to person; one could simply inhale miasmas or be exposed to the sunlight and get sick. In other instances, the interaction between the French and the *indigènes*, aboard the ships for example, was viewed as one of the causes of diseases, because it resulted in frequent disputes, which created "moral emotions" leading to diseases.[22] "All kinds of excesses" were also contributing factors to disease.[23]

The contagionists, on the other hand, contended that yellow fever was caused by a kind of "poison" or "germ" which could be transmitted through contacts with infected buildings and objects such as clothes, furniture, merchandises, mail, partially dry cow skins, leather, wool, and feathers, irrespective of the local sanitary conditions. They believed that yellow fever was not locally produced; instead, it was imported to the Saint-Louis from Sierra Leone, Gambia, Portuguese Guinea or

Grand-Bassam via Gorée and Dakar,[24] but it found in the local "natural insalubrities of the climate" a fertile ground for its spread and lethality.[25] One of the leading proponents of the contagionist school of thought was Bérenger-Féraud who explained the itinerary of the "germ" of the 1867 yellow fever epidemic. According to his view, in November 1866, yellow fever was brought from Sierra Leone to Rufisque, where it killed a local resident, Tachon, whose infected furniture were sold without being fumigated, thus spreading the germ and making victims. In April 1867, another resident, Remaury, who was in contact with the furniture and bed sheets used by the 1866 victims, brought the disease from Rufisque to Gorée, where he died. His room was not disinfected. On 30 July, Weber, a merchant, died in the room previously occupied by Remaury. On 3 August, Marechal, who had visited Weber, died in Gorée; the ship "*L'Etoile*" carried passengers from Gorée to Saint-Louis. So having established the chain of transmission of the "germ," Bérenger-Féraud concluded that those two fatal cases started the new epidemic in Gorée and the passengers aboard "*L'Etoile*," both the convalescents but "carriers" of disease, became "the very appropriate vehicles for the transmission."

The disease started in the southern residential quarter of the city; and one of *Devès and Chaumette* company's employees, who had gone into infected southern quarter to nurse his sick relative, brought the disease back and spread it in the northern part of the city where he lived; and from there the disease spread in the colony. Bérenger-Féraud rejected the explanations given by officials in Saint-Louis that incriminated the "*Furet*," a ship that anchored in Saint-Louis at a later date in August.[26] However, he failed to identify the mode of transmission of the "germ," either through clothes or other objects in the employee's possession.

Bérenger-Féraud adopted the same approach in his search of the origin and spread of the 1878 epidemic. On 14 July, he explained, physician Massola performed an autopsy on prosecutor Batut who had died from yellow fever in Gorée. The next day, he went to Saint-Louis aboard the "*Cygne*" where he stayed four days without opening his luggage, and then went to Bakel. On 29 July, he opened his luggage with the help of Brigadier Duval and put on the jacket he had on the day of the autopsy. On 16 August, Brigadier Duval died from yellow fever, which spread and killed seven out of eight Europeans in the garrison. On 6 September, the "*Espadon*," containing the "germ" of yellow fever, navigated from Bakel to Saint-Louis; one passenger, physician Dalmas, who provided first care to his colleague Massola and was hospitalized on September 9, died four days later and contaminated the city. On 11 September, 371 soldiers from Gorée, Dakar, and Saint-Louis were dispatched to the Upper Senegal; 180 troops succumbed to yellow fever.[27] In this case, too, the "germ" was transmitted through contacts between people and infected objects, such as the jacket or the ship. Bérenger-Féraud saw the 1880 epidemic as the continuation of 1878 epidemic: the *tirailleurs* (African troops), who, in March 1880, disinfected and destroyed the bed sheets from the recent epidemic that were in a closed storehouse, became sick and spread the disease.

Thus, from his personal observations, Bérenger-Féraud saw a link between the introduction and spread of yellow fever and the rapid movement of people and goods. He contended that the primary cause of yellow fever in West Africa was importation from the Caribbean and the American East Coast, and that the environmental and climatic factors and individual physiological and pathological factors were the "secondary causes."[28] Indeed, by the 1870s, it took the *Compagnie des Messageries Maritimes* twenty-five days to go from Bordeaux to Buenos Aries via Lisbon–Dakar–Rio de Janeiro–Montevideo, and twenty-seven days to return to Bordeaux via Bahia and Pernambuco. The Senegalese port-cities of Saint-Louis and Gorée until then, and Dakar, thereafter, became central places in an Atlantic trading network, which extended north to Bordeaux, Marseille, Liverpool, Genes, Havre, Dunkerque, Gottenburg, and Hamburg; west to Guadeloupe, Martinique, Brazil, Argentine and the United States; south to Bathrust, Freetown, Conakry, Grand-Bassam and beyond; and east to upper Senegal.[29]

It should be stressed that the two schools of thought—localists and contagionists—were not incompatible and did not form two permanent groups. There were probably few concerned localists and contagionists but the majority of health officials found both explanations complementary.[30] The 1881 yellow fever epidemic provides a good example of how the medical authorities combined the two theories. Indeed, the Hygiene Commission, in its attempt in July to explain the causes of the epidemic, identified three causes of the outbreak: the stagnant water in the city streets following each rain, the unsanitary houses where deaths were reported during the 1878 epidemic, and the location of the main hospital (Military Hospital) in the middle of the city where it constituted a "focus of infection" during the outbreaks of epidemics. However, the members of the Hygiene and Public Sanitation Council pointed to the three cemeteries surrounding the city as constituting a "permanent danger" to the city's health. The "infectious emanations" coming from the cemeteries, they believed, were likely to aggravate the sanitary situation.[31] But it was physician Merlaut Martialis, Chief Medical Officer, who provided a more detailed account of what he believed were the main causes of the epidemic, drawing on both explanations. He argued that "the imported germs, which had caused previous epidemics and remained dormant for a more or less long time, were able to receive, thanks to the influence of favorable atmospheric and social conditions for their ability to reproduce, all their destructive (or harmful) properties." The evidence in support of his hypothesis came from the first seven reported cases:

> The seven first cases, which preceded the epidemic period, occurred in different parts of the city in rooms where previous yellow fever patients had

stayed...each house playing the role of focus of infection for its neighbors...the infectious germ simply dormant and waiting to rise and manifest itself again with great virulence thanks to favorable circumstances of temperature, humidity, electricity...(and) receptive terrain, etc.[32]

In this post-epidemic report, physician Martialis combined the two schools of thought concerning the response to the threats posed by yellow fever. Thus, the sequence of transmission of the diseases was as follows:

smells/miasmas → humans (localism, anti-contagionism);
infected objects → humans (contagionism)

In other words, there were two instances of contamination: 1) one could inhale the "bad odor" or miasma and get sick; and 2) one could catch the disease by touching an infected object.

Also, when yellow fever next struck Senegal again in 1900, the medical authorities attributed the origin of the epidemic to the importation through a commercial port (Gorée until 1881 and Dakar thereafter) but they believed that the damaged local environmental sanitary conditions facilitated its spread. The outbreak of 1900 provides good evidence for this assertion. Governor General Chaudié explained that

the epidemic seemed stopped at the beginning of the month and all quarantine measures were lifted on 7 June. But on 11 June, following the first rains and the telluric emanations that they provoked, three new cases were reported in Dakar. Two died almost immediately. The third case died only three days after his hospitalization. This third victim was Bishop Buléon, bishop of Cariopolis, apostolic vicaire of Senegambia.[33]

Frustrated that the epidemic was given more importance than it deserved, he went on to argue that 'it was the social status of the deceased that provoked high emotion not only in Senegal but also in France; it also helps explain the exaggeration accorded to the news about the recrudescence of the epidemic.'[34] But he cautioned that 'the rains of the soon-approaching *hivernage* will reveal the true character of the malignancy of the affliction and will help appreciate the degree of contamination of the colony.'[35] Here again, climatic factors played a predominant role in the etiology of yellow fever. Thus, until late nineteenth century, the factors of disease causation, including humidity, brusque variations of temperature in tropical regions, spoiled foodstuffs, overcrowding, unsanitary housing, lack of exercise, and contagion, remained unchanged, as chapter 4 will show.[36] It should be emphasized that the assumptions of racial immunity to disease held by physicians had serious implications in terms of health care delivery.

The evidence suggests that physicians in Senegal correctly observed that most epidemic diseases emerged and spread during the hot season, with its heat and rainfall. But they found that the ways in which the miasmatic theory articulated the correlation between yellow fever and the change in the average temperature during the rainy season was sufficient and did not require additional verification even when yellow fever became a serious health threat. This is the reason why they continued to explain the irregular lapses between epidemics and the re-emergence of yellow fever in terms of bureaucratic effectiveness or deficiencies in the implementation of sanitary legislation and the natives' immunity, and not as evidence of weaknesses or contradictions of the theory. French physicians in Senegal looked for evidence that could explain each new outbreak and rejected particular observations that could falsify the theory, such as the rainy season, the multiplication of mosquitoes and other insects and yellow fever. But, as time went on, it was the recorded high mortality rates among the Europeans that led to the scrutiny of the theory in terms of both logical and empirical criteria. Issues of acclimatization and racial immunity continued to dominate the medical thinking of physicians.

## Yellow Fever Mortality

The available mortality statistics, drawn from the minutes of various health boards' meetings, are inadequate because they recorded only the deaths that occurred in the military and civilian hospitals, isolation camps, and aboard ships. These official statistics reflect the city's stratification and class division. The privileged groups—officers, the "assimilated," European troops, and Senegalese troops—had access to medical services; the "assimilated" included civil servants, employees from local and municipal services and businesses, and self-employed upper- and middle-class individuals and their family members. It was easy for the administration to keep track of their life events. They were buried in the Catholic cemetery. The destitute, who represented only an insignificant fraction of the patients, received health care free of charge in charity dispensaries and the civilian hospital. But the figures related to bed occupancy at the main hospital suggest that the majority of city residents—the *indigènes* and the floating population—had a limited access to health care, even after 1897, when the cost per in-patient and out-patient day was reduced by a third, or 8.56 francs per day per case.[37] Many cases of illness among them were never seen by French doctors, only by local healers. They were buried in the Muslim cemetery far away from European eyes. Thus, the official statistics of disease mortality are incomplete concerning class-specific cases.

## The 1867 Yellow Fever Epidemic

In 1867 yellow fever struck the city at the end of August and spread rapidly, provoking a general panic and overwhelming the response capacity of the colonial authorities and the available health resources. When the disease subsided at the end of October, it had killed around 300 Europeans, including seven Catholic Sisters, who had cared for the sick, and hundreds of natives. Sixty-four patients had recovered from the disease.[38] The epidemic stimulated the administration to put in place the framework of a public health organization by creating one Council of Hygiene and Public Salubrity (CHPS) in Saint-Louis, chaired by *Médecin en Chef* Julien-Henri Rulland, and another in Gorée.[39] The Council recommended the measures that were consistent with the medical thinking of the time, including cleanliness, improving ventilation, water supply, the construction of public toilets, the disinfection of public and private buildings, and the immediate closure of the cemetery in Sor for at least seven years, and the opening of two new cemeteries—one for Catholics in Sor and another for Muslims north of Ndar-Toute.[40] Another board of health, the Sanitary Commission, presided over by *Ordonnateur Tredos*, imposed a quarantine lasting seven to fifteen days to ships dropping anchor at the ports of Saint-Louis and Gorée with an unclean bill of health. By the beginning of the rainy season and fearing the possibility of another deadly epidemic crisis, the CHPS went one step further and imposed a preventive "quarantine of observation" of five days to ships arriving in Saint-Louis from the southern West African coast between Sangomar and Salne, the presumed source of infection, where there were no French Consuls to monitor the sanitary situation, with a clean bill of health.[41] So the 1860s inaugurated for Saint-Louis four decades of uncertainty and probably the most dramatic period in the medical history of the city. Indeed, a cholera epidemic ravaged the city in November 1868 and within a month it had killed 1,112 *indigènes*—5,000 according to missionary sources[42]—and 92 Europeans, including Governor Pinet-Laprade, out of a population of 20,000, as chapter 3 shows.[43] Ten years had passed without another epidemic, and then came the 1878 yellow fever.

## The 1878 Yellow Fever Epidemic

The conjuncture in Senegal in the 1870s was dominated by wars and famine in the countryside. Crowding and starvation provided the virus with new host populations who did not develop immunity against yellow fever or who had lost their immunity. Urban growth had created vast problems of sewerage, waste removal, housing, and water supply, and increased the quantity of breeding places for the mosquitoes. In Saint-Louis, a city of 16,000 residents, the policy of forced removals of the urban poor from the city center had continued between 1871 and 1873 to include Saint-Paul Street in the north and Saint-Joseph Street in the south; the declared objective was to reduce overcrowding and filth. But the environmental and social conditions in the extreme north and south of the city and in the slum areas were favorable for the spread of disease. The French colonial administration was in a precarious situation. It had abandoned Cayor (Louga and Mekhe) in Senegal, and Grand-Bassam and Assinie in Côte d'Ivoire. Gabon was reduced to a simple entrepot for charcoal. Thus, little attention was paid to disease surveillance.

The 1878 yellow fever was part of the pandemic that affected most Atlantic societies from Philadelphia and Memphis to Rio de Janeiro and Freetown as a result of globalization of trade and travel. It arrived from Brazil brought by *Le Niger* and *L'Orénique* in May to Dakar. The boat was not quarantined.[44] The epidemic was declared in Gorée in mid-July and in Saint-Louis in September, only after the infection spread from the *indigènes* to the Europeans. Despite the control measures adopted by the authorities,[45] yellow fever spread in the city and, by December, it had killed 652 Europeans out of a European population of 1,300, including 14 Sisters of the Congregation of Saint-Joseph de Cluny who cared for the sick, and an undetermined number of middle-class members and *indigènes*.

## The 1880-81 Yellow Fever Epidemic

Yellow fever re-emerged in 1880 and 1881, killing respectively 80 and 503 among the European residents, including the newly arrived Governor De Lanneau.[46] The disease was now seen as endemic to Senegal.

Table 2.1. Yellow Fever Mortality in Saint-Louis in 1881

| LOCATION | DEATHS |
| --- | --- |
| Military hospital | 269 |
| City and hospice | 109 |
| Makana isolation camp | 10 |
| Bop-Diarra isolation camp | 13 |
| Aboard ships or camps for crews | 93 |
| Aboard the "Tannesi" in the river | 9 |
| Total | 503 |

*Sources:* ANS/H32/AOF/14, "Epidémie de fièvre jaune à Saint-Louis (23 Juillet–19 Novembre 1881)," by the Chief Medical Officer, January 13, 1882, 10.

Table 2.2. Recapitulation: Official Mortality from Yellow Fever and Cholera, 1867–1881

| YEAR | YELLOW FEVER | CHOLERA | TOTAL POPULATION |
|---|---|---|---|
| 1867 | 300 | – | 15,000 |
| 1868–1869 | – | 1,204 | 15,480 |
| 1878 | 652 | – | 16,000 |
| 1880 | 80 | – | 15,900 |
| 1881 | 503 | – | 15,300 |

*Sources:* see Table 2.1

## The 1900 Yellow Fever Epidemic

The return of yellow fever posed a serious challenge to the physicians' capacity, accumulated over the past two decades, to influence all decisions in the colony. First, physicians had problems posing a correct diagnosis of the disease. Second, they constructed yellow fever as the 'White man's disease' that would subside only *'faute d'aliment'* ('for lack of food'), that is, with the repatriation of all the Europeans, on whom it was supposedly feeding. These two clinical medical problems led to delays in the official declaration of the epidemic, uncertainty on its departure, and contradictory decisions to impose, to lift and to re-impose quarantines.

Although cases of 'fever' were observed since mid-April in Dakar and Rufisque among the civilian population, physicians were unable to identify the disease as yellow fever, thus sending mixed signals to state officials. In one confidential cablegram to the minister of colonies, Governor General Chaudié explained that 'six suspected cases of a disease with the characteristic similar to yellow fever,' followed by deaths, were observed among the civilian population (in) Dakar' (17 May)[47]; in another he recommended that the sentence 'suspected cases of yellow fever' (19 May) be mentioned on unclean bills of health delivered to ships leaving Dakar. On the basis of these unconfirmed cases, he then imposed quarantine measures prescribed by article 8 of the decree of March 10, 1897 to ships that were in communication with Dakar and Rufisque.[48] Thirty-five suspect Syrian peddlers operating between Saint-Louis and other coastal cities were submitted to medical examination and issued identity cards in order to control their movements[49]; the used clothes they sold were suspected of spreading the 'germs' of the disease.

It is important to note that, unlike during previous epidemics, the authorities in Paris closely monitored the ways in which state officials in Senegal managed the 1900 yellow fever epidemic, and they cabled specific instructions to Governor General

Chaudié concerning sanitary measures to be adopted. Convinced that yellow fever was 'a disease that only struck the Europeans' and inspired by the miasma theory, Kermorgant, General Inspector of Colonial Health Services in Paris, recommended the suspension of affectation of troops and civil servants to Senegal and Soudan, the 'meticulous inspection of garrisons' by the Head of Health Service in Senegal to make sure that there was 'no defective hygienic condition,' the return to France of soldiers in 'precarious health' before the set-in of the 'bad season,'[50] and the 'rigorous inspection of premises where the first yellow fever cases had occurred, as well as the clothes, bed sheets etc.'[51] A 'strict disinfection' was conducted; and the municipal authorities were ordered to rigorously maintain the city clean. It was hoped that sanitary measures applied 'with energy would conjure up a grave danger.'[52]

But while the operations of sanitary police were under way in Senegal and French Soudan, Minister of Colonies Decrais had doubts about the capacity of the colonial administration in French Guinea to effectively deal with the health crisis. "Are you," asked Decrais to the authorities in Conakry in a cablegram, "able to implement against yellow fever, all the provisions prescribed by the decree of sanitary maritime police of 31 March 1897?"[53] Kermorgant requested that the same cablegram be sent to Grand Bassam (Côte d'Ivoire), Cotonou (Dahomey), and Libreville (Gabon).[54] Governors Cousturier of French Guinea and Pascal of Dahomey simply replied that the "measures have been taken to assure (the) prescriptions (of the) decree (of 31 March 1897)."[55] Only Governor Lemaire of Gabon had the courage to admit that the colony did not have 'regulatory means' because the heat chamber or disinfection machine ordered three months earlier had not arrived yet. Instead, he was prepared to apply the provisions of article 60 of the decree of March 31, 1897 to regular ships and to deny access to any other ships.[56]

The evidence here shows that the authorities were well aware of the ongoing gap between the policies, expressed in the sanitary legislation, and the realities of shortcomings on the ground, which helps explain the reasons why the authorities in Paris and Saint-Louis seemed in a state of general panic. One just needs to pay attention to the vocabulary utilized to describe the level of preparedness. Adjectives, such as 'meticulous,' 'rigorous,' 'severe,' and 'energetic' attached to the planned activities of inspection, disinfection, and hygiene, as well as the constant reference to yellow fever as the 'disaster' (*fléau*), testify to the state of mind of the officials.

By early June, yellow fever had subsided, followed by sporadic cases in Dakar, Rufisque, and Gorée, and by low mortality rates. In the absence of new cases in ten days and following the recovery and discharge from the Dakar hospital of three patients, Governor General Chaudié, confident that 'the sanitary situation of (the) colony (was) satisfactory,'[57] on 8 June, lifted all quarantine measures and ordered the delivery of clean bills of health.[58] But, in a reversal of fortune, two days later, 'following a brusque invasion of people and rain,' (the 'first rains and the telluric emanations that they provoked'), three new fatal cases were reported and the sanitary situation in Dakar was again perceived as 'unsanitary,' thus prompting the 'necessary precautions.'[59] What contributed to panic was the fact that one of the two patients was Bishop Buléon, Vicar Apostolic of Senegambia. In a hastily held meeting of the *Conseil Privé*, the authorities made an official declaration of yellow fever epidemic.[60] Soldiers were dispatched to the dissemination camps where tents and huts were erected, while plans were put in motion to repatriate 'those (among them) who were exhausted or did not seem to be able to tolerate the bad season.'[61] No physical contact was allowed between camps; communication between them was possible only through clarion or by hand signals. Only the *indigènes*, believed to be immune to yellow fever, carried water and food. Exposure to sunlight, during the hot hours of the day, was prohibited.[62] Plans for sending soldiers and civil servants to Senegal and Soudan aboard the Messageries Maritimes were canceled.[63] Instead, the general European populace from urban centers was encouraged to find vacant seats aboard the 'Bordeaux' and the 'Saint-Louis' commercial fleets and to leave the colony immediately in order to avoid 'big and irremediable disasters.'[64]

By June 23, Governor General Chaudié reported that 'the sanitary situation of the colony more and more leaves much to be desired.'[65] Twenty-seven cases and 15 deaths were reported until then. Reports from Rufisque indicated 'a sudden deterioration of the sanitary situation' and led to the decision to immediately evacuate the city.[66] As official cablegrams were kept confidential, Saint-Louis' residents learned about yellow fever outbreak through rumors. "There is a rumor in the city," wrote one missionary, "that Rufisque and Gorée are contaminated. After verification, it is absolutely true, indeed!"[67] Bishop Buléon's death provoked an immense effect on public opinion and a great emotional response in Senegal and in France as well. Governor General Chaudié believed that the death of such a great personality had something to do with 'the exaggeration with which has been received, in France, the news of the recrudescence of the epidemic.'[68]

The administration in Paris and Saint-Louis scrambled to find vacant seats on the 'Messageries Maritimes' or commercial fleet, such as the '*Compagnie Chargeurs Réunis*,' the '*Taurus*' owned by the Fraissinet Company and the '*Macina*,' that could evacuate all the candidates for repatriation ('*rapatriables*'), including soldiers at the expiration of their stay and those taking convalescence leave, in order to complete the evacuation by 12 July.[69] Twenty-three deputy officers and 183 corporals and soldiers were repatriated in June. French soldiers in Dakar stayed behind to undergo the routine observation quarantine measures that comprised also the disinfection of not only their luggage but also their personal belongings, including the clothes they wore.[70]

Physicians seemed powerless in the face of growing danger. The explanatory power of the epidemiological theory was rather weak, as the theory could not account for an outbreak of yellow fever disconnected from *hivernage* (rainy season), as well as *hivernage* without yellow fever epidemics. The 'current medical constitution' or the 'irregular' and 'disconcerting' evolution of the epidemic worried state and medical authorities as well as city residents, whose vivid memories of past epidemics left a 'terrifying impression' and who feared the worst but hoped for the best. From the physicians' perspective, only the upcoming *hivernage* 'could reveal the true character of the malignity of the affliction and allow an appreciation of the degree of contamination of the colony.' Only then could the 'constitution' go from being 'worrisome' to being 'alarming.'[71]

The re-emergence of the epidemic in early July in Dakar, with its daily new cases and deaths, created a 'sad situation' for the medical personnel.[72] It was reported that yellow fever killed its victims in a spectacular manner, that is, quick and 'as struck by lightning.'[73] The 'only efficient measure of preservation' that could save European lives was their repatriation to France. Having made sure that repatriation did not pose a security threat for the colony in the light of the 'status of foreign relations,' officials began to prepare for evacuation.[74] More than 200 soldiers from Gorée and Dakar were repatriated in July.[75] Merchants, who were worried about the security of their businesses, received assurance that necessary steps would be taken to protect stocks of merchandise by requiring native troops commanded by native officers, joined by local police, to ensure the protection and security of commercial firms and comptoirs.[76] Yellow fever being constructed as the 'White man's disease,' the expectation was that the Senegalese soldiers were immune to the disease.

Although daily reports described Saint-Louis as unscathed, the evidence shows that by at least mid-July there were yellow fever cases and deaths that were erroneously attributed to 'pernicious malaria.'[77] The confirmation of the outbreak gave way to an epidemic of fear, as the following entry from the missionary journal shows: 'We have nothing to envy to Dakar, we are caught. An infantry soldier came down this evening with all the symptoms of the terrible disaster.'[78] Baillif's death, followed by that of two other Navy infantry soldiers from South Rogniat garrison, gave way to raw emotions.[79]

Again, evacuation plans for all French troops and civil servants were put into execution and the business community as well as those in the liberal professions were urged to leave the colony, while other French residents were strongly encouraged either to leave the city or 'to build indigenous-like dwellings on different points of Senegal where they would find refuge ... and wait until the epidemic loses its infectious properties.' Clearly, the authorities were only preoccupied with the protection of the lives of 'those who, because of their origins, offered a sure pray to the contagion,' not those of the Africans who were believed to be immune to the disease.[80] The governor general recommended the evacuation of the French troops from Senegal[81]; the first evacuation of 500 French soldiers and 100 civilians took place on 2 August in panic, carried by the '*Santa Fé*' and the '*Caravellas*' owned by the *Chargeurs Réunis* Company.[82]

As the disease spread and more deaths were reported in August, fear also spread: "Several new cases are reported from everywhere. One does not know where to look at. It is a general panic!"[83] The decision made shortly afterwards by Governor General Chaudié to leave Saint-Louis for France 'despite extreme weakness,' was a turning point in the development of the epidemic and it impacted on the decisions about public health in Senegal. His precipitated departure took place when the number of cases had increased to reach 35 per day[84] and more deaths were reported. The responsibility for running current affairs fell to Colonel Combes, *Commandant Supérieur des Troupes*, until the arrival of Governor Ballay from Guinea for the interim.[85] Four hundred soldiers had to be repatriated from Dakar on 20 August.[86] So between 7 June and 30 August, a total of 1,543 officers and soldiers were repatriated.[87] Only 'strict necessary' personnel were left in place, that is, the Senegalese soldiers (*Tirailleurs* and *Spahis*) and a few of their French supervisors, who were chosen preferably among those 'seasoned,' that is, those immunized through prior infection.[88] They were dispatched in dissemination camps outside the city. Saint-Louis and Dakar had only simple detachments entirely composed of Senegalese soldiers.[89]

On 1 September Governor General Ballay reported that it 'seems that yellow fever increases in intensity' and that 'a simple passage (of ships) suffices to contaminate.'[90] All communications between Saint-Louis and the trading posts on the Senegal River were prohibited, except for the post service and food, drugs, and ammunition supplies that were done by the Messageries Maritimes.[91] The situation was chaotic. *L'Écho de Paris* reported that "a complete anarchy prevails in Senegal since the hasty departure of M. Chaudié and the shameful flight of some civil servants following the example of their chief; it is unacceptable that the telegraph no longer functions."[92] Although the sanitary situation was declared improved in early October with only two new yellow fever cases reported in Dakar and two in Saint-Louis,[93] a week later, the authorities dispatched all the French personnel of the *regiment des tirailleurs* to the isolation stations of Mpal and Pointe aux Chameaux.[94]

Thus, problems of correct diagnosis still lingered on. As mortality cases continued to be reported toward the end of October 1900, the authorities maintained the travel warning and ban. In a cablegram, Governor General Ballay advised the minister of colonies that

> it is thus still imprudent to come to Senegal where epidemic decreases in intensity for lack of food but where still exist ready to re-emerge live germs. It would be better in the general interest for the commercial firms not to send (their) personnel until further notice otherwise the epidemic will re-

cede only very late (while) requiring upholding of quarantine measures (which) are very annoying for all.[95]

The hesitations in the decision-making process reveal the same problems of diagnosis that physicians faced at the beginning of the epidemic. Even if yellow fever was still mentioned as a major cause of death in official statistics, by mid-October, there was a marked tendency to deny its virulence and prevalence; other labels, such as black-water fever (*fièvre bileuse hématurique*)[96] or malaria (*cachexie palustre*),[97] began to re-appear in daily reports of causes of death.

The sanitary situation of the colony remained a subject of preoccupation for the authorities and its clarification determined the fate of the personnel repatriated ('*rapatriés*'). François Roder, a retired Postal and Telegraphic cadre with experience in Senegal and Tonkin, whose son-in-law was planning to join his post in Senegal, raised pertinent questions in a letter addressed to the minister of colonies, dated November 12, 1900: 'Will an officer leaving for Senegal toward the end of December be authorized to bring his wife and his child and do so without any risks for them? Should one fear an offensive return of the dreadful epidemic for next year?' The responses to these questions had to determine whether Roder and his wife would follow their daughter and her family or not.[98] But by mid-November, despite travel warnings issued by Governor General Ballay, the Bordeaux-based ship owners and *négociants* sent back a great number of their employees on the basis of 'cablegrams reassuring them concerning the sanitary situation in Saint-Louis and Rufisque.' Other individuals not associated with commerce also returned to Senegal.[99] The revelation at the end of November of previously unreported 'numerous suspected cases' of yellow fever in Dakar, Kaolack, Fatick, Foudiongue, Thiès, and Tivaouane and two deaths in Tivaouane and Thiès convinced state officials that, contrary to the belief that the epidemic had departed from most coastal cities but Dakar, the disease continued to spread in urban areas as well as in the countryside and that quarantine measures were still needed against those 'contaminated points.'[100] The '*rapatriés*' were invited to cancel their travel plans until further notice. But commercial firms as well as private citizens continued to pressure the governor general and the minister of colonies to fix the public opinion concerning his timing of the official declaration of the end of the epidemic and the 'probable date of that restart.'[101] Despite the persistence of 'isolated cases,' on 2 December, Governor General Ballay declared the 'complete improvement of the sanitary situation in Senegal,' the end of disinfection in all major urban centers, free circulation of trains, except in places not disinfected yet, free movement of merchandise immediately and of people in ten days on the Senegal River; and he authorized the return of '*rapatriés*.'[102] 'I will see, with pleasure,' he wrote, 'the arrival of Robert deputy administrator.'[103] Thiès was the only center that was kept in 'rigorous isolation.'[104] Né-

*gociants*, civil servants, and the other *rapatriés* were invited to return to Senegal. Thus, during the epidemic crisis and in its wake, sustained efforts, heavily inspired by the environmental paradigm and contagionism, were made to control yellow fever, and the authorities had credited the automatic quarantines for the absence of new outbreaks of yellow fever between 1884 and 1899. The re-emergence of yellow fever in 1900 and the progress made in bacteriology provided an opportunity for a paradigm shift, as chapter 4 will show.

## Responses to Yellow Fever: Panic and Medical Profiling

The responses to the outbreaks of yellow fever ranged from rumors, panic and flight to accusations and protests. Given the uncertainty surrounding yellow fever, rumors of confirmed or unconfirmed outbreak reports in Gorée or Dakar caused panic in Saint-Louis. Often, panic led to flight. In 1867 missionaries reported that "a general panic spread among the inhabitants; families were in consternation; physicians themselves were afraid. The hospital was consigned: only the [Catholic] Brothers and [hospital] employees could get in."[105] The panic provoked by the sound of the clarion was such that the governor ordered on request from the citizens that the official funeral ceremonies for the military officers proceed "as taciturn as possible."[106] In 1878, infected city residents fled the city in panic while incubating the disease and spread it into the countryside. The recurrence of the epidemic generated new waves of fear and panic. Father Deplanches noted on July 23, 1881, "Today, when we least expect it, the city is quarantined; several cases of yellow fever are reported. This sanitary measure causes a great emotion."[107] For more than 100 passengers who were preparing to leave for France via Dakar aboard the "*Castor*," the declaration of an epidemic and the imposition of a quarantine eliminated the last opportunity for escape and condemned them to a certain death, which explains the intensity of the emotion. The perception of Africa as "a mysterious land full of dangers,"[108] greatly contributed to the hysteria.

The epidemic of fear naturally led to suspicion, and accusations and counter-accusations. Indeed, the yellow fever outbreaks led the health authorities to construct the *indigènes* as natural targets for particular diseases and as threats to the health of the Europeans and the "assimilated" groups. Such instance was provided in May 1879 as cases of fever and diarrhea accompanied by vomiting were observed among city residents. The anxiety was justified not only by the coming of the rainy season and the possibility of another epidemic, but also because of an influx into Saint-Louis of hundreds of European troops. The unknown illness provoked an intense debate during a key meeting of May 14, 1879 of the Sanitary Commission, presided over by Pierre Carpot, *Ordonnateur*, who represented the administration. In a report to the governor the day before, physicians had linked the symptoms to the offensive smells from rotting fish coming from Guet-Ndar, and had called on the public health officials to remove the "focus of infection." Pierre Carpot blamed Mayor Gaspard Devès, from one of the leading *métis* families, for tolerating the "general unregulated fishing" in Guet-Ndar and for his unwillingness to implement the recent sanitary measures, which had provisions for a new location for fish smoking at the extreme south of Guet-Ndar in order to minimize the threat to the city's welfare. Standing as "the defender of the material interests of the *indigènes* and those of the city," Gaspard Devès took considerable pain to argue that the fishing industry was not a threat to the city's health and that the *indigènes* had the right to dry fish and earn a living. He reminded the participants that even during the tenure of Governor Pinet Laprade (1865–1869) no workable solution could be found to the issue, and that the only viable alternative chosen by the Municipal Council in 1878 was to prohibit fish smoking during the rainy season. He presented contrary evidence indicating that symptoms were sporadic and were sometimes observed among the *indigènes* who ate fish without enough spices and hot pepper during the cold season. He concluded that the causes of the ailments—known as *n'diank*—could be found elsewhere. But where Devès focused on similarities—the same symptoms affecting the indigenes—Dr. Pierre-Adolphe Doué, Acting Chief Medical Officer, saw the essential "differences" between the Europeans and the *indigènes*. His thesis was that "different diets" and "different living conditions" could not possibly result in similar symptoms. He reported that three officers who had recently visited Guet-Ndar contracted the illness while crossing the Servatius Bridge on their way back to the city-island. In response, Devès ruled out the suggestion that the fishing industry would be the "unique cause" of the illness. But a vote at the end of the debate on the issue of "fish smoking as a danger for the sanitary situation of Saint-Louis" and "what to do about it," resulted in a majority in favor of an immediate ban and every year starting on May 1.[109] The authorities believed that the decision would protect the city against "the return of the terrible epidemic" that ravaged the city in 1867 given that the odor of the fish "could develop the germs of the disease." But the decision provoked agitation among the fishermen who saw it as another bureaucratic form of harassment.[110] One of the consequences of the ban was the exodus of adult fishermen in search of work as *laptots* (canoe pilots) or *traitants* along the Senegal River.

In French thinking, the *indigènes* had disqualified themselves from sanitary citizenship by refusing to adopt French cultural values and ideas of "progress," and were therefore situated outside modernity and "civilization." They were "different," "Other," and the social difference was located in their usages, habits, lifestyles, and sexuality. According to the officials, the *indigènes* facilitated the spreading of the diseases by damaging the environmental sanitary conditions. The *indigènes* were seen throwing all the rubbish and dirt into the streets and public places and along the docks, which made an unbearable stink. Many streets leading to the Senegal River on both sides of the island were dead-ends because of the accumulation of the waste matter, including human waste. They urinated on the streets, which explained why certain streets corners smell terribly ("*les odeurs les plus nauséabondes*").[111] Their women were seen littering and bathing naked in public, and these practices were viewed as the result of "self-negligence" and "the absence of sentiments that distinguish people who respect themselves," despite the claim made a decade earlier by Hamet Ndiaye Anne, *Cadi* (judge) of the Muslim Tribunal, that the "old habits have ended."[112] All waste matters accumulated in the northwestern part of the island were disgusting to see.[113] Garbage, filth, bad smells and the succession of outbreaks challenged the basic assumptions about the interaction between the French and the *indigènes*. It was an irony that Hamet died of "fever" on May 15, 1879, the day following the intense meeting of the Sanitary Commission.[114]

The missionaries' response to yellow fever epidemics and the epidemic of fear was framed within the moral paradigm. From their perspective, the outbreaks of yellow fever were punishments from God. Yellow fever struck the city at the time when, from the Church's perspective, people had lost all human respect, compassion, as well as religious devotion; the culture of the city was corrupted. City authorities and many other members of the elite had exhibited a great indifference concerning religious practices, especially at Christmas.[115] God had used yellow fever not only as a way of satisfying His justice but also as a sign of His mercifulness, an opportunity given to the sinners to go back to the sacraments. During the 1881 epidemic, missionaries observed that

> At the hospital, all repented (two or three refused)....The majority of them saw in the first symptoms of the disease a warning from God and promised that, if they recovered, their first visit would be made at the Church and the confessional booth.[116]

During the epidemics, the Church organized processions at the Sor cathedral and said prayers to the Virgin Mary in order to conjure yellow fever. The Catholic Sisters worked overtime providing care and relief to the sick at the hospital and other emergency stations. The Fathers and Brothers of Ploermel visited the patients day and night and relentlessly administered the last rite sacrament; they witnessed remarkable conversions and those who were exposed to mosquito bites during the nights fell sick and succumbed to the disease.[117] Even after the germ theory the missionaries never departed from their moral paradigm.

During the 1900 yellow fever epidemic, if the *indigènes* as a group were considered the source of outbreaks of disease, the authorities, in their zeal to protect the city against the new outbreak, targeted some groups it suspected of spreading the disease. Single women, in particular, constituted convenient scapegoats. They had all the characteristics of a target: they were visible, available, and vulnerable.[118] The archival records identify them as "submissive girls" (*"filles soumises"*), "public girls" (*"filles publiques"*), and "immoral Negroes" (*"négresses de mauvaise vie"*). Some were mulattos and others were Blacks. Four single women, working for the Sazzarin family on Parquet Street as domestic servants and who had friendly rapports with the crew members of the ship "*Saint Kilda*," were blamed for the spread of the 1900 yellow fever. A local publisher and notable, Cornu, affirmed that the crews brought the "germs" of yellow fever and passed them along to the women, who in turn contaminated the soldiers who became the first victims of the disease. Another source of infection, he believed, was a small house located at the end of the same street where other "immoral Negroes" lived and were visited by the troops. He felt that it was his duty to report these "facts" to the authorities.[119] The Police Chief, J. Avrial, confirmed these allegations and recommended that Mrs. Sarrazin's house be disinfected.[120] Thus the authorities considered single women saturated with sexuality and, thus, a threat to city's health.[121]

The evidence suggests that the sanitation syndrome, described by M. Swanson in South Africa, was at work in Senegal as well. It referred to dealing with urban race relations in the imagery of infection and epidemic disease.[122] In Saint-Louis the yellow fever epidemic gave way to the epidemic of fear and suspicion and to the epidemic of stigmatization of the *indigènes*.[123] Guet-Ndar, the most populous *faubourg*, was now seen as the focus of infection and the "space of death," to borrow Michael Taussig's expression,[124] where people, fish, beaches, and even the air were infected and contagious. "Civilized" persons, or sanitary citizens, who went there returned sick. It was the "heart of darkness," to use a familiar expression.

## Strategies to Address Yellow Fever Threats

### Treatment

Yellow fever exposed the limits of Western medicine in the nineteenth century. One of the best descriptions of the treatment of yellow fever available in the surviving archival materials was provided by A. Beziat, the local representative of the Maurel and Prom Company, in a letter to Captain Butez, commandant of the "*Tamezi*," who had lost the majority of his European crews during a trip to the upper Senegal River. Beziat's young brother was successfully treated in 1881 by physician Palmade. The medical treatment of yellow fever available included the following: isolating the patient from other patients in a room with ventilation; providing absolute rest: no noise was allowed around him/her; purging with lemonade Roge, or 50 gr. of sulfate de soude; applying *energetic* frictions with vinegar to kidneys and legs to combat chill; providing little *cold* drink and very light tea so as not to fatigue the stomach; provoking abundant sweating; giving a bath with *séné* the day after purging; providing lukewarm water and three teaspoons of olive oil to facilitate daily bowel movement; and putting *cool* compresses on the patient's head as long as he/she had fever. Patients who survived were also put on a special diet including cold chicken soup for three or four days, an egg, and a combination of chicken or beefsteak for the next week or so before being discharged after recovery.[125]

This description shows the precautions taken to minimize the negative effects of overcrowding and lack of ventilation in the military hospital as well as the importance attached to food in fighting the disease. The silence in Beziat's letter about the technique used to provoke sweating implies that it was part of the therapeutics routinely used to alleviate suffering. Indeed, it is established that therapeutics available to the Hippocratics continued to be used in the nineteenth century, including sudorifics for inducing sweating, febrifuges for combating fever, bloodletting, cathartics, and emetics. W. E. Bynum, a historian of medicine, suggested that new drugs and techniques were added to the old pharmacopoeia, including the Peruvian bark for intermittent fevers, "ipecacuanha for inducing vomiting; [and] a variety of metallic preparations of arsenic, mercury, and antimony."[126] Unfortunately, physicians were unable to save many lives with this treatment.

## Preventive Measures

Efforts to control a disease about which little was known and to avoid its recurrence heightened the anxieties that the French felt about Senegal. Techniques used to deal with the yellow fever threat combined both sanitationist and quarantinist approaches. The localists, or sanitarians, favored strategies aimed at improving sanitation and hygiene enforced by serious municipal police, while the contagionists relied mainly on disinfection of mail, merchandise, and suspect or contaminated houses, the isolation of European troops, and a surveillance program that would enforce the sanitary cordons and quarantine regulations applying to ships arriving from contaminated or suspected regions.[127] In practice, the two approaches were combined. But when these measures revealed ineffective in the face of a growing danger, the authorities opted for the repatriation to France of non-essential personnel.

During the 1878–1881 epidemics, the CHPS members[128] devoted the available health resources to enforcing the quarantine and sanitary cordon regulations, issuing health passes for travelers and goods, setting up isolation camps, lazarettos, and ambulances, cleaning the city streets, regulating the cemeteries, and disinfecting suspect or contaminated houses and objects that had been in contact with the victims.[129] Early on, officials were concerned by the close proximity of the cemeteries that were perceived as an important source of "infectious miasmas" and a "permanent danger" to city health. To minimize the danger, they prohibited burying the corpses less than 35 to 40 meters from the limits of the city and *faubourgs*, or periurban villages. Consequently, they immediately closed down both the Old Catholic Cemetery situated at the entrance of the Faidherbe Bridge in Sor as well as the old Muslim cemetery located south of Guet-Ndar for at least seven years. They opened a new Catholic Cemetery in Sor east of Bouetville at the end of the Rue de Paris with the capacity of 2,181 tombs placed under the authority of the mayor, who delivered authorizations for burial, and a new Muslim cemetery located two kilometers north of Ndar-Toute with the capacity of 5,300 tombs. The location of the new cemeteries was chosen so that the winds would not carry the deadly miasmas through the city. A gravedigger was hired for each cemetery—with the annual salary of 800 francs for the Catholic cemetery and 400 francs for the Muslim cemetery—to pay scrupulously close attention to the depth of the tombs and to maintain the fence.[130] These measures concerned some conditions under which yellow fever could flourish, that is, inadequate sanitation.

The maritime quarantine and the sanitary cordons on land were considered the most effective tools in combating what the authorities referred to as the "pestilential exotic diseases." This expression was used by the lawmakers to refer, early on, to cholera, yellow fever and plague; later on, it applied to other diseases, including typhoid fever, smallpox, diphtheria, scarlet fever, and dysentery. Building an effective response capacity required the following: 1) isolation camps or lazarettos to be used for the quarantine of the passengers and the disinfection of merchandise, other objects, and of the ship itself, and 2) sanitary stations specially arranged in order to accommodate the boats owned by Africans coming from a contaminated region. The lazarettos had to be equipped with an infirmary, a certain number of separate barracks that could accommodate the sick and the suspects, separate stores for merchandise and other objects to be disinfected or already disinfected, safe drinking water, a sewage system, and a machine for disinfection.

Two lazarettos were established at Bop-Diara, south of Saint-Louis on the Langue de Barbarie, a small portion of land separating the Senegal River from the Atlantic Ocean, for large steamships and at Bop-Nkior, north of Saint-Louis on Thionq Island, for barges and small boats arriving in Saint-Louis from the area between Bakel and Podor on the Senegal River.[131] Later, a third quarantine station opened on Babagueye Island south of Saint-Louis to disinfect the merchandise imported from Europe through Dakar to Saint-Louis and the trading centers along the Senegal and Middle Niger Rivers. In these lazarettos, Navy health officers were put in charge of the special operations of sanitary inspection of the passengers and crews and for signs of communicable diseases. The presence of a

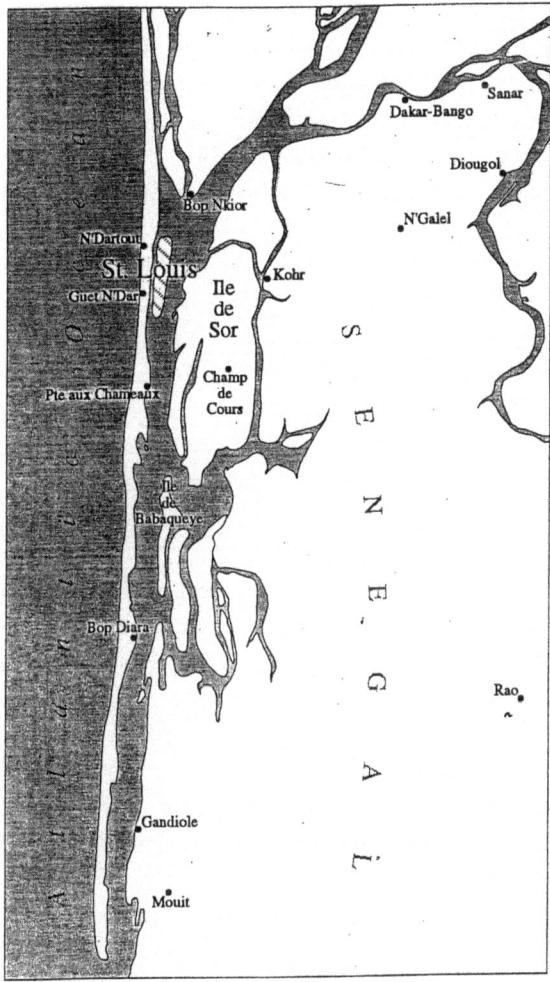

*Map 3. St. Louis and Isolation Camps*

physician aboard postal ships or any ships carrying 100 or more Europeans for more than 48 hours was required. Upon arrival in a port and before communicating with people in the port, any ship not considered "suspect" was visited by the sanitary authority and the captain interrogated under oath on health matters.

If the ship was suspect, then it was submitted to a special sanitary inspection, including, depending on circumstances, a medical visit to passengers and the crews. A contaminated ship was disinfected, including the items such as clothes, beds, carpets, merchandise originating from contaminated regions, and animals. Other objects susceptible to transmitting contagious diseases, such as animal skins, were simply destroyed. A quarantine of observation was imposed upon any ship that had left a country ravaged by yellow fever or plague less than seven days earlier, and a country ravaged by cholera at least nine days earlier. The duration of the quarantine was fixed at five days for the Africans and fifteen days for the Europeans. The underlying mistaken assumption was that Africans and *métis* were immunized against yellow fever. The quarantine was fifteen days for merchandise, excepting groundnuts and gum that were fumigated before embarkation and shipped directly to Saint-Louis without transit through the quarantine stations.[132] Individuals had to pay fees to cover the expenses engaged by the local service for food and maintenance estimated at 7fr50c for officers and first class passengers, 3fr50c for under-officers and economy-class passengers, 1fr25c for corporals and deck passengers, half of the fees for children under 12, free stay for infants, 1fr00c for the disinfection of 100 kilograms of merchandise and passengers' luggage.[133] Owners of ships and canoes were required to provide the "disinfectant substances" needed to disinfect the ships, cargo, and the luggage for the crews and passengers.[134]

A maritime quarantine was supported by a sanitary cordon monitoring all contacts by land that was placed near two bridges giving access to Saint-Louis: Leybar and Khor, where sanitary agents were placed to prevent suspected persons from communicating with Saint-Louis. The sanitary cordon was protected by a task force designated by the *Commandant Supérieur* of troops. The early sanitary cordon was very strict in the sense that it interrupted food supply and trade and isolated the city from its region and the outside world. Pressure and protests led the authorities to extend it to include the post of Lampsar and the Diaudune Bridge, but without allowing people, animals or goods free access to the city. Later, the sanitary cordon became selective; it was no longer imposed on the pedestrians and their local supplies or on the horsemen without merchandise. The main targets were the caravans and individuals carrying imported merchandise and exported products.[135]

Table 2.3. The Quarantines and *Cordons Sanitaires* in Saint-Louis

| YEAR | 1ST CLINICAL CASE | QUARANTINE | CORDON SANITAIRE | LAST CLINICAL CASE | Q. LIFTED |
|---|---|---|---|---|---|
| 1867a | 14 Aug | ? Aug | - | 31 Oct | 10 Feb. 1868 |
| 1878 | 9 Sept | 27 July | 28 July | 17 Dec | 7 Jan.1879 |
| 1880b | ? July | 24 July | - | 13 Ja | 23 Feb. 1881 |
| 1881c | 26 June | 23 July | - | - | ? Dec. 1881 |

Sources: ACSE, *Bulletin Général*, 6.42 (1867), 183–84; ACSE, *Bulletin Général*, 12 (1881–1883), 348–49; *Bulletin Général*, 12, 349–53; ANS/H24/Senegal, minutes of the Sanitary Commission meeting, May 17, 1899; ANS/H41/AOF/120, decree no. 216 of May 20,1901; ANS, *Bulletin Administratif*, decree of June 10, 1900; decree of June 18,1900.

After the train service was established in 1885, Kelle station was chosen as the terminal point for the trains, where all the operations of disinfection were to take place. Regular trains were replaced by sanitary trains operating four days a week between Saint-Louis and Dakar, which transported the non-essential personnel to be repatriated to France.

The panic was such that the bureaucrats who were in charge of combating the epidemic could face severe sanctions in the case of failure. For example, the death penalty was requested for a sanitary agent, a state ship agent or a private company "who would, officially, in a telegram, a certificate, a report, a declaration or an evidence, alter or distort the facts, so as to compromise the public health...if a pestilential invasion followed,"[136] and for agents who would communicate with contaminated regions; life prison and between 200 and 2,000 francs in fees for communicating with countries under the regime qualified as "suspect"; from one to ten years of prison and from 100 to 10,000 francs in fees for a prohibited communication with places, people or things being quarantined.[137] The death penalty was also requested for any individual working in a "sanitary cordon" or any person assigned to the quarantine that would abandon his post or violate the instructions. If the disease did not spread, such an agent would face forced labor and fines between 10,000 and 20,000 francs. Other sanctions could apply to various cases concerning the refusal to implement sanitary measures or the neglect of duties. Thus, the obligation for a ship, no matter its origin, to show its bill of health upon arrival in a port of a French colony or a country under French protectorate was motivated by the need to protect public health.

Another strategy aimed at stopping the spread of the epidemics focused on disinfections or fumigation, with sulfuric or carbolic acid, of postal mail and packages, luggage and merchandise carried by trains and ships, suspect furniture, clothes, public and private buildings and unsanitary houses where deaths were previously reported. The disinfections and painting with lime of unsanitary houses, in particular, was done under the surveillance of the police chief.[138]

Protecting Europeans' lives at the time when territorial expansion in West Africa was high on the administration's agenda also required their isolation from the perceived focus of infection. During the epidemic the military and the civilian hospitals became too small for too many patients. So the surplus patients and the convalescents were transferred to Dakar-Bango, a small camp located outside the city, to Pointe-aux-Chameaux on the Langue de Barbarie, and to a small farmhouse in Sor that was made available to the administration by Mayor Gaspard Devès.[139] European and native troops were dispatched in isolation camps outside the city. Early camps were mobile and moved daily to different locations, such as Leybar, Lampsar, and Ker-Mandoube-Kary. It was Governor De Lanneau who designated several sites that would become regular isolation camps: Pointe-aux-

Chameaux (Navy infantry, Moroccan workers), Lampsar and Makana (artillery), Richard Toll (Spahis), Bop-Diarra and Mauit (*Tirailleurs*).[140] Later, Bop-Nkior, Baba Gueye and Gandiole were added to the list. The main camps could accommodate more than 500 troops. The officers and technicians were confined aboard ships such as *Le Gaston* and *Herminie*.[141] However, for financial reasons the authorities were unable to fully equip the isolation camps and make them permanent. Another preventive measure aimed at protecting the health and lives of European troops led to the evacuation, following the intervention of Deputé Gasconi and some notables such as Louis Descemet and Léon d'Erneville, of the entire garrison of Spahis that was viewed as the hotbed of the 1878 epidemic because of the high mortality rates recorded in its ranks, and their relocation in Pointe Nord where a new garrison was built.[142]

Given the high mortality rates observed during the great outbreaks of yellow fever, the repatriation of the non-essential personnel became the standard practice before the beginning of the rainy season. Observers noted that in early June "a great number of people from Saint-Louis" left for France.[143] Indeed, contingency plans for repatriation were elaborated and ready for execution on a short notice. French troops and civil servants working in Saint-Louis were divided into three groups according to the degree of "seasoning": 1) essential personnel, 2) personnel to be dispatched in isolation camps outside the city, and 3) the personnel to be repatriated. Repatriation was automatic once an epidemic was declared, but because of the irregular intervals between the epidemics, the authorities put in place procedures to grant "regular" repatriation even in the absence of an epidemic. Europeans were eligible for repatriation for three reasons: administrative leave (fatigue, heat, rain), medical leave (convalescence), and permission for sickness.[144] The authorities had hoped the contingency plans would be efficient and would save European and troops' lives without generating unnecessary high costs. But, as we will see later (chapter 4), it was the intervention of the central administration in Paris that helped sustain such expensive plans in the long run. Yellow fever became a serious embarrassment for the officials in the light of their discourse on the Civilizing Mission.

## Power and Knowledge: The Health Service

The re-emergence of deadly epidemic diseases in Senegal enhanced the status of the medical personnel (Navy physicians) in the eyes of the administration officials and made the chief of the Health Service a key player in the decision-making process. Since the 1825–1827 organization of public health in the colonies, Navy physicians, surgeons, and pharmacists operated under the supervision of the *Ordon-*

*nateur*. With the reorganization of the colonial Health Service in 1880,[145] the chief of the Senegal Health Service (*Directeur de la Santé*), along with his colleagues in Guyane, French establishments in India, Cochinchine, and *Nouvelle Calédonie*, gained autonomy and was directly accountable to the governor for two years before being placed under the supervision of the Interior Director in 1882, whose prerogatives extended to include the administration of the sanitary police. Because of the absence in Senegal of civilian physicians with sufficient established technical competence to serve as Chief Medical Officer, the prerogatives of the Chief of Health Service were assumed by the Navy *Médecin en Chef*.[146] But many among the administration officials were uneasy with the new bureaucratic structure. The members of the *Conseil Privé*, well aware that the prerogatives of the Health Service were "the most serious prerogatives given that they are intimately linked to public Safety," decided, during a critical meeting held on February 8–9, 1883, to concentrate all matters related to the sanitary police in the sole hands of the Chief Medical Officer, accountable to the governor, without sharing them with the Interior Director.[147]

The decision to provide autonomy to the Health Service was inspired by the lessons learned from the practice in the Antilles and Reunion, where the Chief Medical Officer was accountable, even on medical matters, to the Interior Director. Such a situation of dependence, so the argument went, insulated the *Médecin en Chef* from 'constant communication with the enemy' and made it difficult for him to impose a sustained quarantine in the light of the interests of commerce and industry. The members of the *Conseil Privé* (Private Council) preferred, instead, the situation in France, where since 1881 the *Médecin en Chef* was accountable to the Minister, instead of the *Préfet maritime*; and the Minister was represented in the colony by the Governor, not the Interior Director. The proposed bureaucratic arrangement empowered the Chief Medical Officer who would, in case of an epidemic, take over the decision-making process.[148]

A decade later, as the questions treated in Senegal, may they be military, commercial or administrative—such as the construction of a garrison, hospital, school, relations with the Middle and Upper Senegal River, or a military expedition—increasingly raised issues related to hygiene or sanitary police, pressure mounted to admit the *Médecin en Chef* into the Defense Council (*Conseil de Défense*) and Private Council (*Conseil Privé*).[149] Indeed, the decree of January 28, 1892 appointed the Chief Medical Officer as a member of the *Conseil Privé* [150] with deliberative voice. In a report to Mr. Carnot, President of the Republic, the Minister of Commerce, Industry and Colonies, Jules Roche, highlighted the importance of the Health Office in the following terms:

> This service has become so important in this (colonial) possession, where most of the affairs are linked, by the force of circumstances, to the issues of

hygiene and sanitary police, that it is indispensable to modify this situation (by admitting the Head of Health Office in the *Conseil Privé*).[151]

The decree had important implications in terms of social control. The physician's role in shaping and maintaining the colonial order sharply increased. It would not be an exaggeration to argue that Senegal had then entered a phase of "medicalization" of colonial affairs. Government officials had to seek the opinion of the Chief Medical Officer (*Médecin en Chef*) in various instances, such as the creation of an administrative post, especially concerning its location, its spatial layout, and the arrangement of buildings; the official declaration of an epidemic or its end; the appropriate control measures; and the assessment of human responsibilities in the emergence or spread of the outbreak.[152]

The emergence and re-emergence of yellow fever epidemics and the increasing mortality affecting top leadership as well as the workforce had serious political impact on the daily management of colonial affairs, the continuity of administrative and political personnel, and institutional memory. The evidence suggests that the perception of differential susceptibility to disease could have contributed to the empowerment of the General Council, dominated by the *métis* who were believed to have acquired immunity against yellow fever. The constant disease threat could also have been instrumental in the career choice made by some students from the leading *métis* families. The career of Charles Carpot is a good case in point. Instead of the predictable *cursus honorum* leading from the law school in Paris to the career in politics, as his brother François Carpot and others did, Charles went to medical school and returned to Senegal to practice medicine. As chapter 4 will show, Dr. Charles Carpot would play a major role during the 1900 yellow fever epidemic not only in the construction of the epidemic itself but also in dealing with the subsequent political crisis and the crisis of confidence.

## Conclusion

The growth of Saint-Louis in the framework of international trade created serious health problems that highlighted the inadequacies of public health policies. High mortality from yellow fever epidemics underlined the anxieties over the viability of Saint-Louis, as a colonial settlement, and Senegal, as a colony. Physicians played an important role in the construction of Senegal as an "unhealthy" colony and the constant reformulation of the fears of the tropics centered on the concepts of climatic determinism of disease, racial susceptibility, levels of seasoning, and repatriation policy. They also contributed to the construction of the *indigènes* as "carriers" of the infectious agents. The colonial administration used the available resources to combat an enemy that remained elusive and re-emerged when unexpected. The *Médecin en Chef* who led the "war" against yellow fever was awarded with the leadership of the newly reorganized Health Service accountable to the governor, not the Interior Director. He set up the framework for understanding the mechanisms of disease causation and increasingly gave his opinions on all colonial matters, not just health issues. Physicians also contributed to the cultivation of racial difference through the construction of yellow fever as the "White man's disease" in opposition to cholera that was constructed as the "Black man's disease." This story will be the object of the next chapter.

## Notes

1. Khaled Bloom, *The Mississippi Valley's Great Yellow Fever Epidemic of 1878* (Baton Rouge: Louisiana State University Press, 1993), 4.
2. P. Curtin, *Death by Migration. Europe's Encounter with the Tropical World in the Nineteenth Century* (Cambridge: Cambridge University Press, 1989), xiv, 130; Idem, *Disease and Empire: The Health of European Troops in the Conquest of Africa* (Cambridge, U.K.; New York: Cambridge University Press, 1998), 10; Donald B. Cooper and Kenneth F. Kipple, 'Yellow fever,' in Kenneth F. Kipple (ed.), *The Cambridge World History of Human Diseases* (New York, 1993), 1101; for a critique of Curtin, see Bruce S. Fetter, 'History and Health Science,' 427. H. Haguenot, *Mémoire sur les Dangers des inhumations* (1771), and Vicq d'Azyr, *Essai sur les lieux et les dangers des sépultures* (1775).
3. Michael B. Oldstone, *Viruses, Plagues, and History*, 16; Stephen S. Hall, 'Billions of powerful weapons to choose from,' in Maya Pines (ed.), *Arousing the Fury of the Immune System: New Ways to Boost the Body's Defense*, 6–29.
4. Dr. Marel Leger, "Les Epidémies de fièvre jaune au Sénégal: aperçu chronologique,' in *Compte Rendu et Commissions de l'Académie des Sciences Coloniales* 10 (1928–29), 395–404; Donald B. Cooper and Kenneth F. Kipple, 'Yellow fever,' 1102–03.
5. Francois Delaporte, *The History of Yellow Fever: An Essay on the Birth of Tropical Medicine* (Cambridge, MA, 1991).
6. W. Biddle, *A Field Guide to Germs* (New York: H. Holt, 1995); M. Oldstone, *Viruses, Plagues, and History* (New York: Oxford U. Press, 1998).
7. Lawrence A. Sawchuk, "Deconstructing an Epidemic: Cholera in Gibraltar," in D. Ann Herring and Alan C. Swedlund (eds.), *Plagues and Epidemics: Infected Spaces Past and Present* (New York: Berg, 2010), 95.
8. J. L. A. Webb Jr., "Malaria in Early Tropical Africa," in Paula Viterbo and Kalala Ngalamulume (eds.), *Health and Medicine in Africa: Multidisciplinary Perspectives* (Berlin: LIT Verlag, 2010; East Lansing: Michigan State University Press, 2010), 127.
9. William E. Connolly, *Identity/Difference: Democratic Negotiations of Political Paradox* (Ithaca, NY: Cornell University Press, 1991), 40.
10. ANFSOM/Senegal/XI27a, Salva to Governor, no date listed.

11. Alain Corbin, *Le Miasme et la jonquille: L'Odorat et l'imaginaire social, 18e–19e siecles* (Paris: Editions Aubier Montaigne, 1982).
12. Donald Reid, *Paris Sewers and Sewermen: Realities and Representations* (Cambridge, MA: Harvard University Press, 1991), 10–17.
13. The same approach was used by the British physicians in India; see David Arnold, *Colonizing the Body: State Medicine and Epidemic Disease in Nineteenth-Century India* (Berkeley: University of California Press, 1993), especially chapter 1.
14. Bérenger-Féraud, *Traité clinique des maladies des Européens au Sénégal*. Vol. 1 (Paris: Adrien De La Haye, 1875), viii.
15. ANS, *Moniteur du Senegal et Dependances*, 1873, 126–27.
16. *Journal de la Communauté, 1904–1951*, entry for May 1905.
17. Archives Nationales du Senegal (ANS) H20/Senegal, records of the Hygiene and Public Sanitation Council meeting, 10–11 January 1868.
18. ANS/H20/Senegal, records of the Hygiene and Public Sanitation Council meeting, 10–11 January 1868.
19. ANS/H20/Senegal, records of the Hygiene and Public Sanitation Council meeting, 10–11 January 1868.
20. ANS/H 25/Senegal, minute (record) of the meeting held by the Municipal Commission, 30 July 1900.
21. ANS/H32/AOF/29, Gouvernement du Senegal, *Bulletin Sanitaire*, 11 January 1881.
22. ANS/H11/AOF/5, Rapport de Mr. Doublet, Capitaine du navire l'Etoile, 1860, 41–42.
23. ANS/H32/AOF/29, Gouvernement du Senegal, *Bulletin Sanitaire*, 11 January 1881.
24. ANS/H20/Senegal, report of the Hygienic and Public Sanitation Council meeting, 10–11 January 1868; for more on contagion and infection, see Peter Baldwin, *Contagion and the State in Europe, 1830–1930* (New York, 1999), 3–5.
25. A. Du Mazet, 'Note sur l'assainissement du Sénégal,' *Moniteur du Sénégal*, 21 February 1882, 31–33.
26. Berenger-Feraud, *De la fièvre jaune au Sénégal* (Paris, 1974), 131–36.
27. Berenger-Feraud, *Traité théorique*, 157
28. Bérenger-Feraud, *Traité théorique*, x, 444–59, 460.
29. ANS, *Moniteur du Sénégal et Dépendances*, January 1877, 16; June 1878, 132, 137–38, 142–43, 147–48, 154–55.
30. ANS/H32/AOF/14, 'Epidemie de fièvre jaune à Saint-Louis (23 juillet–19 novembre),' by Dr. Martialis, 3–4, 6.
31. ANS/H2/Senegal, minutes of the Hygiene and Public Sanitation meeting, 7 September 1881.
32. ANS/H32/AOF/14, "Epidemie de fièvre jaune à Saint-Louis (23 juillet–19 novembre 1881)," by Dr. Merlaut Martialis, 3–4, 6.
33. ANFSOM/FM/SEN/XI/50, Governor General Chaudié to minister of colonies, June 26, 1900.
34. Ibid.
35. Ibid.
36. ANS/H39/AOF/9, Chief Medical Officer to governor general of French West Africa, no. 268 of February 28, 1896.
37. ACSE, *Bulletin de la Congrégation des Soeurs de Saint-Joseph de Cluny*, LVI Dec. 1899, 669; ANS/H?/AOF, "Rapport sur l'application au Sénégal du règlement du 10 mars 1897 sur le

fonctionnement des hôpitaux, by Dr. Primet, 21 May 1897, 7; ANS/H48/AOF/1, "Mission sanitaire du Sénégal. Hygiène Militaire (Febr.–March 1901)," by Dr. Grall, 56; ANS/2G/AOF/13.26, Annual Report, 1913, 62.
38. ACSE, *Bulletin Général* 6.42 (1866–67), 183–4; ANS/H2/III, letter to the ordonnateur, Oct. 9, 1867.
39. ANS, *BOS*, decree of Oct. 25, 1867, 252. The Saint-Louis council included Dr. Rulland, *médecin en chef*, Mayor Dumont, Dr. Beaussier, Dr. Bancal, the Chief Pharmacist, the hospital administrator, the Head of Service de Génie, the commander of the Spahis garrison, Granges (*négoçiant*), and Lautier (property owner); see ANS/Senegal/H20, minutes of the CHPS meeting, Jan. 10–11, 1868. In 1873, an ordinance modified the composition of the council to take into account such changes as the creation of the Interior Service and the Chamber of Commerce; these services were then represented in the council; see *MSD*, 1873, 61–2.
40. ANS/Senegal/H20, minutes of the CHPS meeting, Jan. 10–11, 1868, 8–10; a series of ordinances closed down the old Catholic cemetery located at the entrance of the Faidherbe Bridge and the old Muslim cemetery located south of Guet-Ndar, opened a new Catholic cemetery in Sor, east of Bouetville. See ANS, *Bull. Adm. of Senegal*, ordinances of July 18, 1868, 196–98; see also *Moniteur du Sénégal et Dépendances*, July 21, 1868, 115–17.
41. ANS/Senegal/H20, minutes of the CHPS meeting, June 1, 1868.
42. ACSE, *Bulletin Général*, t. 6 (1868–9), 864.
43. ANFSOM, Série géographique, Sénégal, XI 30, chief medical officer, Rulland, to governor, 3 Jan. 1869.
44. ANFSOM, ministry of agriculture and commerce to the ministry of Navy, 16 May 1878. Cited in Série géogr. Sénégal XI 34, minutes of the Navy Health Superior Council meeting of 30 Jan. 1879.
45. ANS/H28/AOF/2, governor's decision no. 15 of 2 July 1878; ANS/H28/AOF/4, decision of governor, no. 17 of 27 July 1878; ANS/H28/AOF/5, decree no. 16 of 28 July 1878; ANS/H28/AOF/37, *ordonnateur* to governor, 6 Aug. 1878.
46. ACSE, *Journal de la Communauté*, entry for 4 Aug. 1881.
47. ANFCAOM/FM/SEN/XI/50A, Cablegram, Chaudié to minister of colonies, May 17, 1900.
48. ANFCAOM/FM/SEN/XI/50B, Cablegram, Chaudié to minister of colonies, May 19, 1900.
49. ANS/H41/AOF/3, decision of governor general, 22 May 1900; see also ANS/H41/6, Chief Medical Officer to governor general, 27 May 1900.
50. ANFCAOM/FM/SEN/XI/50A, Report to minister, by Kermorgant, May 21, 1900.
51. ANFCAOM/FM/SEN/XI/50A, Minister of colonies to governor general of French West Africa, May 21, 1900.
52. ANFCAOM/FM/SEN/XI/50B, Cablegram, Chaudié to minister of colonies, May 22, 1900.
53. ANFCAOM/FM/SEN/XI/50A, Minister of Colonies Decrais to Guinée française, May 23, 1900.
54. ANFCAOM/FM/SEN/XI/50A, Kermorgant to Decrais, May 22, 1900.
55. ANFCAOM/FM/SEN/XI/50A, Cousturier to minister of colonies, May 24, 1900; ibid., Pascal to minister of colonies, May 26, 1900.
56. ANFCAOM/FM/SEN/XI/50A, Governor Lemaire to minister of colonies, May 26, 1900.
57. ANFCAOM/FM/SEN/XI/50A, Chaudié to minister of colonies, June 4, 1900.
58. ANFCAOM/FM/SEN/XI/50A, Cablegram, Governor General Chaudié to minister of colonies, June 8, 1900.

59. ANFCAOM/FM/SEN/XI/50A, Cablegram, Governor General Chaudié to minister of colonies, June 10, 1900; ibid., 11 June. See also ANFCAOM/FM/SEN/XI/50, Governor general to minister of colonies, June 25, 1900.
60. ANFCAOM/FM/SEN/XI/50B, Mission sanitaire du Sénégal. Rapport d'ensemble, by physician Grall, June 9, 1901, 1–3.
61. ANFCAOM/FM/SEN/XI/50A, Bureau Militaire, 'Note sur l'épidémie de fièvre jaune du Sénégal,' Febr. 13, 1901.
62. ANFCAOM/FM/SEN/XI/50A, Colonel Rabier (14th Regiment, Goree-Dakar) to colonel commandant la place, June 1, 1900.
63. ANFCAOM/FM/SEN/XI/50A, minister of colonies to minister of navy, 11 June 1900,
64. ANFCAOM/FM/SEN/XI/50, Cablegram, Chaudié to minister of colonies, June 11, 1900.
65. ANFCAOM/FM/SEN/XI/50, Cablegram Chaudié to minister of colonies, June 23, 1900.
66. ANFCAOM/FM/SEN/XI/50, June 22, 1900.
67. ACSE, *Journal de la Communauté de Saint-Louis*, entry for 23 June 1900, 384.
68. ANFCAOM/FM/SEN/XI/50, Governor General Chaudié to minister of colonies, June 25, 1900.
69. ANFCAOM/FM/SEN/XI/50A, Minister of colonies, 3e Direction 2e Bureau, 'Note pour le Bureau Militaire,' June 30, 1900. Six hundreds soldiers evacuated by the 'Vauban' arrived in Bordeaux on 12 July; see cablegram no. 899b of July 12, 1900.
70. ANFCAOM/FM/SEN/XI/50A, Minister of colonies to governor general, July 2, 1900.
71. ANFCAOM/FM/SEN/XI/50, governor general to minister of colonies, June 25, 1900; ANFCAOM/FM/SEN/XI/50, governor general to minister of colonies, July 26, 1900.
72. ANFCAOM/FM/SEN/XI/50A, Cablegram, governor general to minister of colonies, July 7, 1900.
73. ANFCAOM/FM/SEN/XI/50A, Cablegram, Chaudié to minister of colonies, July 27, 1900.
74. ANFCAOM/FM/SEN/XI/50A, governor general to minister of colonies, July 11, 1900.
75. ANFCAOM/FM/SEN/XI/50A, Bureau Militaire, 'Note sur l'épidémie de fièvre jaune du Sénégal,' Febr. 13, 1901.
76. ANFCAOM/FM/SEN/XI/50, Cablegram, Minister of Colonies Decrais to governor, July 2, 1900 (the cablegram was never sent, the minister thinking that it was not necessary to do so).
77. ANFCAOM/FM/SEN/XI/50A, Cablegram, governor general to minister of colonies, July 17, 1900.
78. ACSE, *Journal de la Communauté de Saint-Louis*, entry for 17 July 1900, 385.
79. ANFCAOM/FM/SEN/XI/50A, Cablegram, governor general to minister of colonies, July 22, 1900.
80. ANFCAOM/FM/SEN/XI/50, Governor General Chaudié to minister of colonies, July 26, 1900. (a 4-page letter). Five hundred soldiers and 100 civilians were repatriated aboard 'Santa Fe' and 'Caravellas,'; see governor general to minister of colonies, July 28, 1900.
81. ANFCAOM/FM/SEN/XI/50A, Navy minister to minister of colonies (Bureau Militaire), August 4, 1900; Chaudié to minister of colonies, August 10, 1900.
82. ANFCAOM/FM/SEN/XI/50, Note pour la 3e Direction (2e Bureau), no. 303 of August 2, 1900. Four hundred soldiers were repatriated on August 8, 1900; see ANFCAOM/FM/SEN/XI/50A, Cablegram, governor general to minister of colonies, August 8, 1900. Four hundred soldiers were to be repatriated on 20 August aboard the 'Campana.' Chaudié noted in the last cablegram that his 'personal health continues to improve.'

83. ACSE, *Journal de Saint-Louis*, entry for 11 Aug. 1900, 386.
84. ANFCAOM/FM/SEN/XI/50A, Cablegram, Comber, governor general a.i. to minister of colonies, August 18, 1900.
85. ANFCAOM/FM/SEN/XI/50, Cablegram, Chaudié to minister of colonies, no. 170 of August 14, 1900; on 4 August, Governor Chaudié informed the minister of colonies that 'I am still extremely tired, I stay in bed.' See cablegram, Chaudié to governor general, 4 August. On August 5, 1900, Minister of Colonies Decrais urged Governor General Chaudié to take care of himself and to prepare for departure at the first opportunity; see ANFCAOM/FM/SEN/XI/50A, Decrais to governor general, August 5, 1900. On August 9, the Conseil des Ministres in Paris authorized Governor Ballay, on his own insistence, to travel to Saint-Louis; he saw the move as a 'duty'; see ANFCAOM/SEN/XI/50A, Cablegram, minister of colonies to governor, Aug. 9, 1900. In September, Chaudié requested permission to return to Saint-Louis in early October; see Chaudié to minister of colonies, Sept. 16, 1900; his demand was rejected and he was invited to 'take advantage of your stay in France to assure the complete recovery of your health before resuming the exercise of your functions.' See minister of colonies to Chaudié, Sept. 26, 1900.
86. ANFCAOM/FM/SEN/XI/50A, Note pour le Bureau Militaire, Aug. 11, 1900.
87. ANFCAOM/FM/SEN/XI/50A, Bureau Militaire, 'Note sur l'épidémie de fièvre jaune du Senegal,' Febr. 13, 1901.
88. ANFCAOM/FM/SEN/XI/50A, Cablegram, Minister of Colonies Albert Decrais to governor general, Sept. 19, 1900.
89. ANFCAOM/FM/SEN/XI/50A, Bureau Militaire, 'Note sur l'épidémie de fièvre jaune du Sénégal,' Febr. 13, 1901.
90. ANFCAOM/FM/SEN/XI/50A, Cablegram, Ballay to minister of colonies, Sept. 1, 1900.
91. ANFCAOM/FM/SEN/XI/50A, Cablegram, Ballay to minister of colonies, Sept. 5, 1900.
92. *L'Echo de Paris*, Wednesday 5 Sept. 1900, no. 5943, 1.
93. ANFCAOM/FM/SEN/XI/50A, Cablegram, Governor General Ballay to minister of colonies, Oct. 4, 1900.
94. ANFCAOM/FM/SEN/XI/50A, Governor General Ballay to minister of colonies, Oct. 9, 1900.
95. ANFCAOM/FM/SEN/XI/50, Cablegram, Governor Ballay to minister of colonies, no. 234 of October 27, 1900.
96. ANFCAOM/FM/SEN/XI/50A, Idem, Oct. 16, 1900.
97. ANFCAOM/FM/SEN/XI/50A, Idem, Nov. 23, 1900.
98. ANFCAOM/FM/SEN/XI/50A, François Roder to minister of colonies, Nov. 12, 1900.
99. ANFCAOM/FM/SEN/XI/50A, Cablegram, Head Colonial Service in Bordeaux to minister of colonies, Nov. 16, 1900; see also cablegram, Decrais to governor of Senegal, Nov. 17, 1900.
100. ANFCAOM/FM/SEN/XI/50A, ibid.
101. ANFCAOM/FM/SEN/XI/50A, President of Administration Council, P. de Trar, to minister of colonies, Nov. 27, 1900.
102. ANFCAOM/FM/SEN/XI/50A, Decrais to Sulpicy, Dec. 3, 1900.
103. ANFCAOM/FM/SEN/XI/50A, Ballay to minister of colonies, Dec. 3, 1900.
104. ANFCAOM/FM/SEN/XI/50A, Ballay to minister of colonies, Dec. 11, 1900.
105. ACSE, *Bulletin Général*, 6.42 (1867), 183.
106. ANS/H2/AOF/109, letter to governor, no date provided.
107. ACSE, *Journal de Saint-Louis*, entry for 23 July 1881, 110.

108. ACSE/158/B/Senegal/024, *Soleil*, 5 Nov. 1878.
109. ANS/H22/Senegal, minutes of the Sanitary Commission meeting, 14 May 1879.
110. ANS/3G3/4/319, note by the director of political affairs, May 20, 1879.
111. ANS/2 G/10/26, 68.
112. ANS/H 21/Senegal, minutes of the Special Commission meeting, 2 July 1864; ANS/H20/Senegal, minutes of the CPHS meeting, 10 Jan. 1868; ANS/Senegal/H22, minutes of the Sanitary Commission meeting, May 14, 1879.
113. Archives Municipales de Saint-Louis/2B20, governor to the President of the Municipal Commission, no. 86, 13 Oct. 1881.
114. ACSE, "Journal de la Communauté, 1852–1890," entries for 15–16, 1879, 101.
115. ACSE, *Annales*, 20.
116. ACSE, *Bulletin Général*, 12 (1881–3), 351.
117. ACSE, *Bulletin General*, 6.42 (1868), 186–8.
118. For more information on the characteristics of a target, see Hubert M. Blalock Jr., *Race and Ethnic Relations* (Englewood Cliffs, NJ: Prentice Hall, 1982), 14–15.
119. ANS/H44/234, Cornu to governor general, 21 July 1900.
120. ANS/H45(AOF)/235, J. Avrial to governor general, 21 July 1900.
121. For more on the "hysterization of women's bodies," see Michel Foucault, *The History of Sexuality*, vol. 1. *An Introduction*. Trans. by Robert Hurley (New York: Penguin Books, 1976), 104–5.
122. M. Swanson, 'Sanitation Syndrome Bubonic Plague and Urban Native Policy in the Cape Colony, 1900–1909," *Journal of African History*, XVIII, 3 (1977),' 387.
123. For more on the subject, read Philip Strong, 'Epidemic Psychology: A Model,' *Sociology of Health and Illness* 12.3 (1990), 249–59.
124. Michael Taussig, *Shamanism, Colonialism and the Wild Man: A Study in Terror and Healing* (Chicago: University of Chicago Press, 1987), 3.
125. ANS/H44/AOF/171, A. Beziat to Governor General Chaudié, 14 June 1900; A. Beziat to Captain Butez, 27 Aug. 1881. The emphasis is in the original letter.
126. W. E. Bynum, *Science and the Practice of Medicine in the Nineteenth Century* (New York: Cambridge University Press, 1996), 18.
127. ANS, Moniteur du Sénégal et dépendances, 1867, 643; ANS/H20/Senegal, minutes of the CPHS meeting, 10–11 Jan. 1868. The duration of the preventive quarantine of observation was seven to fifteen days in 1867, fifteen days in 1878, and twenty-one days in 1881 and thereafter.
128. The CHPS members were Mayor Devès (president), Dr. Bourgarel (Navy Chief Medical Officer and Head of the Health Service), Simon (Chief Pharmacist), Descemet (Chamber of Commerce vice-president), Walter (Chief of Pont et Chaussées Service), Dr. Dubouch (civilian physician), Beziat (habitant notable), Crespin (filling in for Head of Interior Service), Veterinarian (absent).
129. ANS, *Moniteur du Sénégal et Dépendances*, ordinance of 30 Sept. 1867, 64; idem, 2 Aug. 1881, 179–80. In 1881, the duration of the quarantine of ships arriving from the south was fifteen days.
130. ANS/H20/Senegal, minutes of the CHPS meeting, 11 Jan. 1868, 8–10; ANS, *Bulletin Administratif du Sénégal*, ordinance of 18 July 1868, 196–98; ANS, *Moniteur*, 21 July 1868, 115–17.
131. ACSE/158/B/Senegal, minutes of the Sanitary Commission meeting, 13 Sept. 1878.
132. ANS, *Moniteur*, 17 Sept. 1878, 203–4.
133. ANS, *Bulletin Administratif du Sénégal*, 1879, 400–401.

134. ACSE/158/B/Senegal, minutes of the Sanitary Commission meeting, 13 Sept. 1878, 203–4.
135. ANS, *Moniteur*, 15 Oct. 1878, 225. After the train service was established in 1885, Kelle station was chosen as the terminal point for the trains, where all the operations of disinfection were to take place. Regular trains were replaced by sanitary trains operating four days a week between Saint-Louis and Dakar, which transported the non-essential personnel to be repatriated to France.
136. ANS/H2, law of 3 March 1822, art. 10.
137. ANS/H2/AOF, law related to the sanitary police, title II, art. 7–14, 3 March 1822.
138. ANS/H32/AOF/10, mayor to governor, 1 Aug. 1881. The laws of 19 Jan., 7 March, and 13 April 1850 on unsanitary housing were promulgated in Senegal on Oct. 6, 1882; see ANS, *Moniteur*, 1882, 182.
139. ANS/H7/AOF/32, governor's decision, 20 Oct. 1878; ANS/H7/AOF/33, governor's ordinance, 22 Oct. 1878.
140. ANS/H32/AOF/36, *Circulaire* of Governor de Lanneau, 29 July 1881.
141. ANS/H32/AOF/36, Circular of governor, 29 July 1881.
142. ANS/L13/51, note from the interior director a.i., Turquet, June 15, 1895; the military hospital was later built on the site where stood the Spahis garrison.
143. ACSE, *Journal de Saint-Louis*, entry for 13 June 1885, 196.
144. ANS/H43/AOF/20, governor general to *Commandant Supérieur des Troupes*, 25 May 1900.
145. ANS, *M.S.D.*, decree of November 28, 1880, 258.
146. ANFSOM, Sér. Géogr., Senegal, XI–28, Report to the President of the Republic, July 29, 1884 concerning the organization of the sanitary regime in Senegal.
147. ANFSOM, Sér. Géogr., Senegal, XI–28, minutes of the *Conseil Privé* meeting, Febr. 10–11, 1883; the decree of Sept. 12, 1882 created the Direction of Interior with, among other tasks, the administration of the sanitary police run by Navy physicians in the absence of civilian physicians in Senegal.
148. Ibid.
149. ANFSOM/Sér. Géogr./Senegal/VII/26, *médecin en chef* to governor, July 6, 1891.
150. ANFSOM/Sér. Géogr./Senegal/VII/25, decree of Febr. 24, 1885 creating the *Conseil Privé* in replacement of the *Conseil d'Administration*; members were Beziat Jean and Raymond Martin; supply councilors were Victor Beymis and Albert Laplène.
151. ANS, *Bulletin Administratif*, 1892, 65–66; see also ANFSOM/Sér. Géogr./Senegal/VII/26.
152. ANS/H9/AOF/77, Head of Health Service to governor general of French West Africa., no. 296 of November 1895

CHAPTER THREE

# The "Black Man's Disease": Cholera and Social Inequality, 1868–1899

The process of city growth in the context of international commerce and travel created the conditions favorable for the outbreak of viral, bacterial, and parasitic diseases. The cholera epidemic, through the underlying disparities in its distribution and consequences, revealed the urban society to itself. Although the connection of Saint-Louis to the world economy made the city especially vulnerable to the fourth cholera pandemic in the 1860s, and the fifth pandemic in the 1880s and 1890s, it was rather the paucity and the unequal distribution of urban amenities, combined with compromised immune systems of the urban poor, that determined the pattern of mortality and morbidity among city dwellers. The solution to the problem of supplying Saint-Louis with clean water was not separated from the structures of social inequality in the city. The emergence and re-emergence of water-borne diseases corroborate Paul Farmer's argument that disparities "are biological in their expression but are largely socially determined."[1] Social responses to cholera varied greatly. They ranged from fear, suspicion, accusations, stigmatization, and medical profiling to the "culture of poverty" explanations, and to calls for conversion to Christianity.

## Origin and Spread of Cholera

It is established that cholera originated from the coastal regions surrounding the Bay of Bengal, both Bangladesh and the Indian subcontinent, where it was endemic and from where it spread in the framework of international trade and travel. There are seven recorded pandemics of cholera since 1817. The first pandemic occurred between 1817 and 1823 when cholera swept through major Indian cities from Calcutta to Madras and Bombay (today's Mumbai) before spreading beyond its borders to Burma, China, Sri-Lanka, the Philippines, Japan, South Pacific, Iraq, Syria, Russia, and Reunion and Mauritius Islands. From Iraq and Syria, cholera entered Europe.[2]

The second pandemic (1829 to 1851) started in Pen-Njab, spread to Pakistan and Afghanistan, and reached Iraq and Persia (Iran) by land; from there, cholera spread across Russia, Austria, Poland, Finland and, through the Baltic port-cities, to England and Ireland in 1832. In the meantime, a major cholera epidemic was reported in 1831 from Mecca, where it produced thousands of victims. Pilgrims returning to their homes spread the disease in Tunisia, Palestine, Syria, and Egypt. From England, cholera entered France through Calais, struck Paris where it produced 18,406 deaths,[3] and crossed into Italy and Spain. Cholera also moved across the Atlantic Ocean in 1832 through ships from England and Ireland. New York City, Philadelphia, and New Orleans were especially hard hit. Cholera reappeared in Paris in 1848, when the city was also embroiled in a political revolution that led to the establishment of the Second Republic, and in London in the summer of 1849. Dr. John Snow was credited with connecting the cholera outbreak to drinking water contaminated with sewage in Broad Street, Golden Square, and nearby streets, and with stopping the Broad Street epidemic.[4] Ships leaving Europe also brought the cholera to Mexico, Nicaragua, Peru, Chili, and Cuba.

The third pandemic took place between 1852 and 1859. Again, it started in India and was brought by sea by pilgrims to Mecca from where it spread to Egypt, Sudan, Madagascar, and Mauritius Island. By land, it followed the same pattern as the second pandemic to spread in Europe. Wars and population movements created favorable conditions for its spread.

The fourth pandemic occurred between 1863 and 1879, also started in India and spread along the same routes as the previous two. Pilgrims from Java brought cholera to Mecca and transformed it into a foyer of infection in 1865. Pilgrims who survived it spread cholera in Europe, North Africa and Abyssinia. Caravans from Morocco brought the disease to Senegal in 1868, as explained below. Cholera cases and deaths were also reported in Guinea Bissau. Protecting Europe from cholera spread by Muslim pilgrims triggered an international conference in Constantinople in 1866 aimed at harmonizing international sanitary regulations. No agreement came out of the conference, however. The same is true concerning the 1874 Vienna conference. It was not until 1877 that European powers instituted three-day quarantine at El Tor for all ships and pilgrims arriving from contaminated countries en route to Mecca.[5]

The fifth pandemic again started in India in 1881 and swept across the globe from 1881 to 1896, following international commerce and travel. New international conferences were organized to deal with the cholera menace in Washington in 1881 and in Rome in 1885 but with no tangible results. But the lethality of the cholera epidemic in India and the vulnerability of the Suez Canal to the spread of cholera led to the adoption at the 1892 Venice conference of sanitary measures informed by the new science of bacteriology. In addition to maintaining the maritime quarantine at El Tor, the new sanitary measures included surveillance of the Suez Canal and the classification of ships into clean, suspect and contaminated and the definition of specific action to be taken in each case. In 1893, cholera again spread to Senegal and the rest of West Africa from Morocco, as the pages below show. The same year, the Dresden Conference added the compulsory notification of cholera cases and the institution of sanitary surveillance to the defensive arsenal. The 1894 Paris Conference regulated the pilgrimage to Mecca in case of infection from Asian pilgrims.[6]

The sixth pandemic from 1899 to 1923 spread from India and followed the same routes as the previous pandemics, at the exception of sub-Saharan Africa that was not affected for reasons that remain unclear. The seventh pandemic spread from 1961 to the present; it started in Indonesia and affected the six continents. Scientists believe that the first six pandemics were caused by the classical biotype V. cholera of the 01 serotype, while the seventh pandemic was caused by V. cholera 01 El Tor.[7] Like in the case of yellow fever and other epidemics, the main sources of cholera epidemics have been human travel and migrations, such as trade caravans, religious pilgrimages, and military campaigns.[8]

## The Ecology of Cholera

Drinking water and eating food contaminated by human waste constituted the second major cause, after the mosquito-borne diseases, of the high incidence of ailments such as diarrhea, dysentery and other gastrointestinal disorders, typhoid fever, and cholera in Saint-Louis. Unlike diseases such as yellow fever, smallpox, or malaria, which affected all the segments of the urban population, water-borne diseases in general, and cholera in particular, chose as their victims primarily the urban poor and the under-nourished. As the historian Frank Snowden pointed out, cholera "is an infallible indicator of destitution and squalor, of overcrowded dwellings and sanitary neglect, of defective sewers and unwashed hands, of suspect produce and recycled clothing."[9] Indeed, *Vibrio cholerae*, the bacterium that causes the disease, spreads through fecal contamination of the water supply, the consumption of fish and shellfish contaminated by bilge water from the ships that are dumped into the harbor, contaminated foodstuffs (fruits and vegetables which have been washed with infected water), and infected clothes and linen, especially the bed linen of victims. The bacterium survives better in polluted water than in lakes or rivers; it can survive for up to 20 days in river water, up to 15 days in feces, and a week in ordinary earth dust.[10]

Although the German bacteriologist Robert Koch has been credited with the discovery of the *Vibrio cholerae*, the cholera bacillus, in a Calcutta tank in 1884,[11] the reassessment of the available evidence led some scholars to argue that it was rather the Italian scientist Filipo Pacini, who was the first person to discover the cholera bacillus in 1854. The proponents of this alternative interpretation contended that Koch was given credit for reasons that had to do with the national prestige, the strong state intervention in German society, and "the need for the German empire to begin to impose itself on its various constituent federated kingdoms, principalities, and states."[12] Koch's discovery was based on shaky foundations, and his postulates difficult to prove; it remained in scientific limbo until the mid-1890s, when a medical team confirmed his findings.[13]

The incubation period can vary from four hours to four days. The disease can kill up to 70% of symptomatic patients within three to twelve hours. The symptoms are spectacular and terrifying. According to Charles Briggs,

> The toxin produced by the *Vibrio cholerae* bacteria paralyzes the gut in such a way that intestinal cells secrete water and electolytes, resulting in diarrhea and extremely rapid dehydration. Persons who are acutely symptomatic suddenly begin to expel an unbelievable volume of diarrhea and vomit.... The rapid dehydration leaves cholera patients weak and thirsty, their arms and legs grow cold and clammy, and powerful cramps seem to shrivel their limbs and tie them in knots. The tips of their tongues and their lips turn blue, their eyes sink back into their sockets, and their skin hangs on their bodies.[14]

Therapy in 80% to 90% of cholera cases consists of oral rehydration with ORT solution including sodium chlorine (3.5 gr.), sodium bicarbonate (2.5 gr), potassium chlorine (1.5 gr.), glucose (20 gr), and potable water (1 liter).[15]

Nineteenth-century public health officials distinguished two kinds of cholera based on the manners of propagation of the disease: 1) the "epidemic" cholera, or general city-wide outbreak that is caused by the general fecal contamination of the municipal water supply (the river, the cisterns, and wells), and 2) the "sporadic" cholera that is limited to a specific neighborhood, a particular city street, or a house.[16] Twentieth-century epidemiologists have identified two cycles of transmission of cholera rather than two kinds of cholera: 1) the long cycle that spreads through the contamination of the water supply (delta, lagoon, brackish water, and wells), and 2) the short cycle, whose spread is associated with the movements of people (nomads and refugees) and the concentration of people (festivals, pilgrimage, and funeral processions and other burial ceremonies). Infection in the short cycle occurs through dirty hands and contaminated foods and silverware.[17] The difference between the two classifications seems small.

Saint-Louis experienced both modes of propagation of cholera. "Epidemic" cholera or the short cycle was reported respectively in 1868 as an extension of the fourth pandemic (1863–1874), apparently imported to the city by the caravans of traders, and in 1893 by sea in the framework of international travel and trade that spread the fifth pandemic (1881–1896). The epidemic found its place in the official records because of its scope and the danger it represented for the colonial order. "Sporadic" cholera or the long cycle affected mostly the urban poor and, because of its social character, it often went unnoticed by the authorities. The unknown disease, which had its hotbed in Saint-Louis' slums of Guet-Ndar, Ndar-Toute and Sor, and which generated a heated debate in May 1879 (see chapter 2), was in fact a local outbreak that can be classified as a "sporadic" or long cycle cholera. It was called *n'diank*, and regularly visited the urban poor, among whom it disabled or killed hundreds. Some adult survivors could develop protective immunity for up to five years.[18]

Among the factors that contributed to the spread of cholera among city residents were poor sanitation, poor housing, contaminated water supply, and lack of sewerage, which often resulted from public health policy deficiencies. The discussion on yellow fever epidemics in the previous chapter touched on the sanitary conditions that predisposed the city to epidemics, as well as on the administration officials' slum clearance policy. There is no need to revisit the issue here. Rather, the emphasis will be on the relationship between overcrowding, filth, water supply and sewerage on the one hand, and cholera and other water-borne diseases on the other. John Snow, physician and epidemiologist who studied the outbreak of cholera in London in the mid-nineteenth century, explained the ways in which cholera spread among people who shared the same overcrowded dwelling in the following terms:

> The bed linen nearly always becomes wetted by the cholera evacuations, and as these are devoid of the usual colour and odour, the hands of persons waiting on the patient become soiled without their knowing it; and unless these persons are scrupulously cleanly in their habits, and wash their hands before taking food, they must accidentally swallow some of the excretion, and leave some on the food they handle or prepare, which has to be eaten by the rest of the family, who, amongst the working classes, often have to take their meals in the sick room: hence the thousands of instances in which, amongst this class of the population, a case of cholera in one member of the family is followed by other cases.[19]

Snow's description of the mode of transmission of cholera matches eyewitness accounts of the crowded and dirty living conditions in the extreme north and south of Saint-Louis, and the peri-urban villages of Guet-Ndar, Ndar-Toute and Sor in the second half of the nineteenth century. Overcrowding, poor housing, and lack of sanitary amenities exposed family members and neighbors to the feces of the sufferers or immune carriers, which highlighted the serious problems of waste removal that the city experienced.

## The Sewerage Problems

Saint-Louis, described by European visitors in the nineteenth century as the most beautiful European city in West Africa, was actually an unhealthy place. Water supply and sewerage were crucial issues, especially with the transformation of the trading post into a colonial port-city, the result of the subsequent population growth. Private houses had no latrine pits. To rid themselves of their excrement, the elite utilized vases called "*marseillaises*" that the indigenous domestic servants emptied either on the banks of the Senegal River or on the beach twice a day in the morning and the evening. The urban poor used the river and the beaches as cesspools.

The lack of water and sewage systems was part of the larger problem of waste removal in the city, including the household wastewater, garbage, animal waste, and even rain water on the city streets during the *hivernage* months, especially between July 15 and September 15. Plans for building the city's sanitary infrastructures were discussed but abandoned because of either lack of financial support or they were not viable. Some engineers favored the system called the "*tout à l'égout*" (mains drainage) which disposed of all the wastes and rain water at once; others argued in favor of "*le système séparé*" (a separated system) which would evacuate only the kitchen and human wastes, leaving the rain water to be disposed of by other means. The "mains drainage" required building a large drainage system that would be used in full capacity only between July and September, but would have high maintenance cost in terms of the volume of water needed to keep it clean and avoid fermentation. The "separate system" had the advantage of eliminating the matters believed to present "the most immediate danger" to city health. House owners would be required to extend the sewer service pipes to their houses themselves, while the administration would build public toilets in the peri-urban slums. But neither plan was technically, politically, economically, or socially viable. City dwellers were left vulnerable to the "infected ponds" and "fermented substance-targets" present in and around their homes[20] that were responsible for the diseases of the gastrointestinal system. Let us now turn to the problems of water supply in Saint-Louis.

Table 3.1. Changing Level of Salt in the Senegal River, Saint-Louis 1851–1852 (per 100 gr.)

| JUNE 1 | JULY 1 | AUGUST 1 | SEPT 1 | OCT 1 | NOV 1 |
|---|---|---|---|---|---|
| 0.2530 | 0.0618 | 0.0640 | 0.0013 | 0.0009 | 0.0016 |
| DEC. 1 | JAN. 1 | FEBR. 1 | MARCH 1 | APRIL 1 | MAY 1 |
| 0.4915 | 1.0906 | 2.0570 | 2.0325 | 2.0302 | 0.6250 |

Source: Bérenger-Féraud, "Etudes sur la Sénégambie," *Moniteur du Sénégal*, 1873, 111–12.

## Water Supply Problems

Saint-Louis residents enjoyed an abundance of water only four months per year during the *hivernage* (or rainy) season between June and November, with abundant rains particularly between mid-July and mid-September. City residents obtained fresh water from the Senegal River and the rain water collected in cisterns. Government cisterns met the needs of the civilian hospital, the Roignat garrisons, and the *Génie* workshops. Since only 100 out of 14,900 city dwellers had private cisterns, the majority of residents had to rely on unhygienic river water. But during the eight months of dry season, between December and mid-July, water supply was a serious problem,[21] since the river water turned brackish and became polluted. Table 3.1 shows the transformation of the quality of water, as measured in a study conducted by Dr. Audibert, chief pharmacist at the military hospital in Saint-Louis.

The table shows that the water in the Senegal River around Saint-Louis became fresh in mid-July and could be used safely through December, which explains why Governor Valière fixed, in 1870, the date of January 4 for the beginning of distribution of water from a government cistern to civil servants and the military.[22]

During the period of water shortages, the well-to-do utilized the water they had stored in the cisterns during the rainy season, while the administration put in service a ship water-tank, which carried water from upstream at the Kassack flood plain tributary of the Senegal River to the city. Kassack was located near Makana village, 16 kilometers east of Saint-Louis. Ships, such as "*La Sénégambie*" (in the 1860s–1870s) and "*L'Akba*" (in the 1880s–1890s), represented life and hope for many city residents. The water was distributed as indicated below. Animals owned by the administration and used for transportation received water from this source: 10 to 12 liters of water per day per mule and 12 to 14 liters per horse.[23]

The exclusive distribution of water to the officials and other administration employees underlined an important dimension of health inequality. The working class and the urban poor were left to fend for themselves.

## Table 3.2. Distribution of Water to Government Employees, Saint-Louis, 1873

| CIVIL SERVANTS AND TROOPS | QUANTITY OF WATER/DAY |
|---|---|
| Head of administer & Comptroller Chiefs of corps & services | 12 *livres* |
| Milit. officers & assimilated | 8 *livres* |
| Low-ranked officials | 5 *livres* |
| Low-ranked mil. officers, corporals, troops, sailors, other agents | 4 *livres* |
| Family members of officials | 4 *livres*/person |
| Administration employees | 2 *livres* |
| Total: | 35 *livres* |

*Source:* Bérénger-Féraud, Dr., "Etudes sur la Sénégambie," *Moniteur du Sénégal et Dépendances*, 14 Jan. 1873, 9.

They had to rely on ground water tapped by small wells that they dug outside the city in the dunes in Sor or in the Pointe de Barbarie south of the city. The water drawn from these wells was poor in quantity and quality, and the wells rapidly filled with salty water. The poor continued to bath and wash their clothes and utensils in the polluted river water; they were known for drinking the "bad water."[24] Yet, it was possible in Saint-Louis to buy water from private individuals, as seen in an advertisement published in the *Moniteur du Sénégal et Dépendances* in 1864:

> NOTICE
>
> FOR SALE: Potable waters for 1 franc a cask, from the cistern owned by Gilbert. Contact Mr. Boudou.[25]

Even when available for sale on the market, like in this case, the cost of water was still out of reach for the urban poor. Civilian prisoners also did not have access to potable water during the dry season. In 1868 Dr. Beaussier, *médecin* 1st class, observed that the water distributed to the prison population was brackish and contained a large amount of salty ocean water. He recommended that all the prisoners, military and civilian, be equally treated, and given water from the cisterns.[26]

By the early 1870s the price of one cask or barrel of water had increased from 12 to 15 francs.[27] But the poor could not afford to buy the water. Some Muslims who could hold a marriage certificate issued by the *Cadi/Tasmir*, the head of the Muslim tribunal, showing that they had only one legitimate wife, were allowed to obtain the water distributed by the administration for ofranc13 cents a liter.[28] How many people qualified to buy the water is not known, but the majority of the poor were excluded for financial reasons.

During the dry season, water shortages and the contamination of water supplies were responsible for many cases of typhoid fever, diarrhea, dysentery, and gastrointestinal disorders.[29] City residents knew "from experience" that during the "season of the winds from the east", the diseases of the intestinal system as well as a variety of respiratory conditions from the bronchitis/pneumonia/influenza group increased.[30]

## Early Efforts to Build Waterworks

Saint-Louis' expanding population led to more waste matter being discharged directly into the river, and to more exposure to infectious agents. The administration was well aware of the fact that water shortages constituted a bottleneck to urban growth. In the 1860s, Governor Léon Faidherbe attempted to build artesian wells in the northern part of the city. The cost of building a well 250 m deep in order to bring water to the surface was estimated at 54,000 francs but the appropriate technology was not available in the colony. The construction of the well was suspended pending the arrival of the new machine from France.[31] The Inspector of the *Service des Ponts et Chaussées* recognized that "the lack of water is one of the principal obstacles to the extension of our domination."[32]

Governor Pinet-Laprade decided to provide the city with a permanent source of clean drinking water by building waterworks. Indeed, in 1866, a year after he took office, the governor erected a barrage on the Kassack flood plain at Lampsar in order to create a man-made reservoir of fresh water measuring 600 m long, 70 m large and 2.5 m depth. The estimated capacity was 8,500,000 m$^3$, which was sufficient to meet city residents' daily needs.[33] But the barrage was destroyed by floods. In the meantime, the working class and the urban poor were "at risk" from the first "epidemic" cholera of 1868–1869.

## The 1868 Cholera Epidemic

The fourth cholera pandemic spread from India and China thanks to the intensification of international travel facilitated by the new means of transportation such as railways, steamships, and the opening of the Suez Canal in 1869. Mecca, the pilgrimage center, rapidly became the central place for the transmission of cholera: 35,000 out of 90,000 pilgrims succumbed to cholera there in 1865. The pandemic had reached some port-cities of North and West Africa, including Saint-Louis and Bissau city in Portuguese Guinea—where it produced thousands of victims—and Central and Southern Africa (Belgian Congo, Tanganyika, Mozambique, and Madagascar).

One year after the yellow fever epidemic struck Saint-Louis, the first four cases of cholera, followed by deaths, appeared in the city on November 24, 1868. Soon, the number of new cases reported at the Military Hospital rose to 164. The situation was very critical and the city was in a state of panic. Two days later, the administration decided to dispatch troops in the camps outside the city: the *Tirailleurs Sénégalais* (African troops) and the Navy artillery were sent to Gandiole, and the Navy infantry and the *Compagnie des Disciplinaires* were stationed at

Table 3.3. Daily Reported Cholera Deaths, Saint-Louis, 1868

| DATE | INDIGÈNES | EUROPEANS | DATE | INDIGÈNES | EUROPEANS |
| --- | --- | --- | --- | --- | --- |
| 27 Nov. | 28 | – | 14 | 8 | 1 |
| 28 | 53 | – | 15 | 8 | 2 |
| 29 | 60 | 3 | 16 | 11 | 1 |
| 30 | 78 | 3 | 17 | 7 | 2 |
| 1 Dec. | 77 | 10 | 18 | 10 | – |
| 2 | 94 | 1 | 19 | 6 | – |
| 3 | 93 | 7 | 20 | 2 | – |
| 4 | 92 | 9 | 21 | 10 | – |
| 5 | 99 | 13 | 22 | 18 | 1 |
| 6 | 54 | 6 | 23 | 11 | 2 |
| 7 | 55 | 8 | 24 | 13 | 1 |
| 8 | 40 | 4 | 25 | 1 | 1 |
| 9 | 27 | 5 | 26 | 14 | – |
| 10 | 36 | 5 | 27 | 2 | 1 |
| 11 | 33 | 1 | 28 | 7 | – |
| 12 | 17 | 3 | 29 | 9 | – |
| 13 | 15 | 2 | 30 | 9 | – |
|  |  |  | 31 | 1 | – |
| Total |  |  |  | 1112 | 92 |

Source: ANS/AOF/H27/s.n., Chief Medical Officer H. Rulland, to governor, "Rapport sur l'epidémie de cholera de Saint-Louis (Sénégal) en 1868," 3 Jan. 1869; see also ANFSOM, Sér. Géogr. Sénégal, XI d. 30, Annexe A.

Lampsar. The *Spahis* were dispatched in the vicinity of Gandiole. They were given instructions not to stay at one location more than one day. The Brothers of Ploermel and the Sisters of Cluny, who ran the school system, sent the students back to their parents. The panic was general and city residents began to flee. Within days about 10,000 people, out of a population of 20,000, had left the city in panic.[34]

The 1867 yellow fever epidemic had provided health officials with some experience in handling an epidemic crisis. So as early as November 29, 1868, they organized an ambulance or temporary shelter for the sick in the southern part of the island, and transformed one ship into a floating hospital. All the new cases, even those from the military hospital, were to be sent there. The governor also ordered the cleaning of the city streets and docks that were covered with trash and filth.

The disease quickly spread throughout the city. It was "a situation of unseen gravity." Doctors had no effective drugs to combat cholera. The treatment consisted of 1 gr of sulfuric ether, 1–9 cc of laudanum, 20–29 cc of gum syrup, 100–109 cc of gum solution, and 1 teaspoon of P.V.A. every twenty minutes. Another form of treatment was also used in a simple formula: to warm the patient's body up in order to stimulate the circulation by using hot tea or hot bricks, and friction of the arms and legs with *eau de vie camphrée* before covering the patient with a blanket.[35] Such a treatment alleviated suffering from the painful muscular cramps that accompanied the disease but it was ineffective because it did not replace the fluids lost in diarrheal

Table 3.4. Distribution of Deaths from Cholera, Saint-Louis, 1868

| LOCATION | EUROPEANS | | | *INDIGÈNES* | | |
|---|---|---|---|---|---|---|
| | CASES | DEATHS | CASE-SPECIFIC MORTALITY (%) | CASES | DEATHS | CASE-SPECIFIC MORTALITY (%) |
| Military hospital | 38 | 16 | 42.10 | 11 | 9 | 81.81 |
| Floating hospital | 28 | 23 | 82.14 | 68 | 45 | 66.17 |
| Ambulance | 9 | 7 | 77.78 | 208 | 170 | 70.97 |
| Lampsar camp | 44 | 40 | 90.90 | – | – | – |
| Gandiole camp | 8 | 6 | 75.00 | 37 | a.40/26 | 70.27 |
| City (St.-Louis) | – | – | – | – | 848 | – |
| Total | c.127 | b.92 | – | c.324 | b.1112 | – |

Source: ANFSOM Sér. géogr. Sénégal XI d.30, Henri Rulland to Governor, January 3, 1869. Cases, deaths, and case-specific mortality are drawn from Annexes A and C.
    a. There is a discrepancy between officially listed deaths in Annexes A (40) and C (26).
    b. The total is from Annexe A.
    c. The total is from Annexe C.

stools. The evidence is scanty concerning other types of therapy used to combat cholera, but it would not be exaggerated to assume that the strategies used against yellow fever, such as purging and bloodletting, were also tried against cholera. In any case, the number of deaths increased rapidly, as the following table shows.

The table shows that by the end of the epidemic, in December 1868, about 1,204 city residents, including 1,112 *indigènes* and 92 Europeans, out of a population of 20,000, had died in Saint-Louis as well as in various isolation camps, as the following table indicates. This was in just 25 days. From eyewitness accounts, the cholera epidemic had surpassed in intensity the yellow fever epidemic which ravaged the city the previous year.[36]

The actual number of deaths was higher than the officially listed deaths. The statistics show that 92 Europeans out of 127 patients admitted at different medical facilities had died; 250 *indigènes* and *métis* out of 324 patients had also died, including 100 African troops. However, the majority of *indigènes* who died in the city, estimated at 848, did not have access to medical care. They were probably abandoned in their huts by their parents and relatives who fled the city in panic. Missionaries' sources put forward the frightening number of 5,000 deaths.[37]

## Figure 3.1. Daily Reported Cholera Deaths, Saint-Louis, 1868

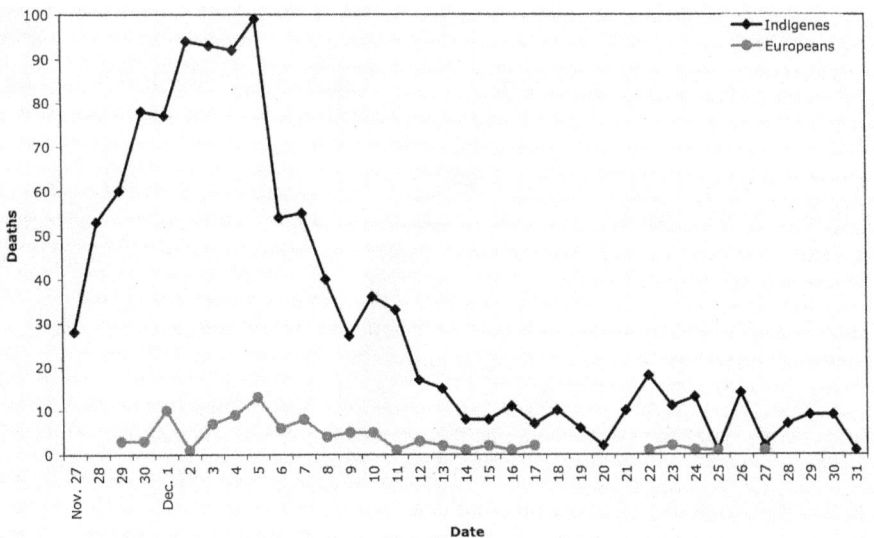

This figure is supported by census data that indicate a population decline passing from 18,840 in 1868 to 11,706 residents in 1869.[38] The data also indicate that cholera had concentrated its attacks among the urban poor, who had no reliable fresh water supplies, and who were more exposed to filth than the elite.

Physician H. Rulland's report to the governor contained only the officially recorded number of cholera and deaths. For since mid-December the mortality rate had considerably declined. Being the most lethal of infectious diseases capable of killing patients within hours, cholera certainly produced more deaths than the report suggested when one takes into account the deaths that occurred incognito in the city or during the flight.

The rural impact of the 1868 cholera epidemic in the Upper Senegal Valley, which formed Saint-Louis' hinterland, is difficult to assess for lack of sources on demographic profile, cases, deaths, and survivorship. French and Senegalese merchants visited the trading posts on the banks of the Senegal River but there was no medical infrastructure or medical personnel to document background information and crisis medical events. It is not exaggerated to argue that cholera did cause extensive mortality and morbidity in fishermen's villages, and in rural communities and trading posts along the Senegal River. In the process, it interrupted economic activities. In any case, on January 11, 1869 the Sanitary Commission lifted the quarantine measures that were taken in order to stop the spread of cholera in the city.

The health authorities' approach to cholera was informed by the miasmatic and contagionist theories. Cholera was seen as the "Black man's disease" par excellence, a disease of the urban poor who lived in unsanitary conditions and had contributed to the environmental pollution. The difference was even more accentuated in the missionary discourse that underlined the religious identity of the majority of cholera victims: they were "Muslims." Cholera was also understood as a contagious disease that spread from person to person even if its mode of transmission remained mysterious. Indeed, the debate about the origin of cholera focused on the Trarza Moors as a "hotbed of infection." It was believed that cholera spread from North Africa, especially Morocco, to Mauritania and Senegal. Travelers had previously observed the symptoms characteristic of cholera, including diarrhea, vomiting, cold, and painful muscular cramps, among the Moors in their homeland in Mauritania. Because many Moors were present in Saint-Louis when the disease struck the city, the conclusion was that they had introduced yellow fever as well as cholera in Saint-Louis.[39] But the missionaries did not share the officials' analysis; they believed that cholera was imported from Timbuktu by the caravans of traders who came to Saint-Louis instead. The difference is minimal anyway since the two claims point to the North African origin of the disease. Thus, Muslim pilgrims and traders were suspected of spreading cholera.

It should be emphasized that the administration considered the Moors an obstacle to the development of trade relations between Saint-Louis and the Walo and Cayor regions located in its immediate hinterland. The Moors had the reputation of being difficult partners in the gum trade, who very often closed down the *escales* or marketplaces along the Senegal River. One of the official reasons for the construction of the bridges between Saint-Louis and the continent in the 1850s was to enable a quick intervention of troops against the Moors, who attacked the local population in Saint-Louis' hinterland around the village of Ndiago and in the north. Thus, the attribution of the responsibility for the outbreak and death of thousands of city residents as well as rural population to the Trarza Moors made them the official enemies. Consequently, their identification as enemies justified the war against them, and probably made other Moor groups vulnerable to harassment and persecution.[40]

Missionaries interpreted the cholera outbreak as a punishment from God and as a divine warning for penitence. Throughout the epidemic, the Catholic Church organized public prayers at the parish as well as processions in order to seek protection from the Saint Virgin, at the Governor's displeasure. At the end of the epidemic, missionaries viewed the high mortality rates among the Muslims as an indication of divine justice. "The Saint Virgin has protected us," they wrote in the journal.[41] Consequently, in the aftermath of the cholera epidemic and in response to the "divine call for conversion," missionaries opened an orphanage for the children from the poor Muslim families who had lost their parents to the disease. It was a golden opportunity for baptism but also a *casus belli* from the Muslims' perspective, who deeply resented the Catholics' encroachment on their holy space.

## The Government Response

The 1868–1869 cholera epidemic came as a warning call to the administration to find a permanent solution to the problem of clean water supply. In early 1869 the director of the *Bureau du Génie* prepared a detailed project for water supply, which was approved by the governor. Water would be drawn from Lampsar and taken to a pumping station at Makhana, and transported through a main aqueduct to a pumping station at Khor in Sor. From there the water would be taken to a high-level reservoir and a water tank in Saint-Louis where it would be stored before being distributed to city residents through smaller water pipes. Each resident would get at least 100 liters of water per day. The reservoir would be built either on the ground previously occupied by the marketplace, or near the church, or on the Pointe du Nord in the middle of the public garden. The cost of the project was estimated at 980,000 francs, and the annual operating expenses were estimated at 19,000 francs. The empty buildings of the old military post at Lampsar would be used to house the maintenance personnel, and to store the equipment and fuel for the operation of the pumps.[42] The project was approved by the *Conseil d'Administration* on June 30, 1869, as cholera struck the city for the second consecutive year.

## The 1869 Cholera Epidemic

Cholera reemerged first in Gambia in early May 1869. It then spread rapidly throughout various towns in Senegal such as Gorée, Dakar, Podor and Dagana. Cholera struck Senegal at the time when the administration was at war with Amadu Sekhu in Cayor. The latter found in cholera an opportunity to recruit a following. The evidence suggests that there was a link between war and the spread of the cholera epidemic in the sense that war provoked the movement of people, caused famine, and exposed the refugees, whose immune system was already weakened by poor diet, to contaminated water and food. Soldiers were probably instrumental in the spread of cholera in the city.

The disease reappeared in Saint-Louis on May 25 when two soldiers who became ill were admitted to the hospital and cured. But by early June the case rate drastically increased, and six to seven deaths were reported daily. News from Bakel, a town located in the Upper Senegal, was alarming: 800 people had died in the town, including the local physician. The beginning of July witnessed a stabilization of the number of cases at seven to eight deaths every day. The majority of the victims were poor city residents, who had little access to clean drinking water and safe food; until then, only three Europeans had died. However, on August 18, Governor Pinet-Laprade (47 years old) succumbed to cholera, followed by the death of Dr. Maurel.[43] They were probably exposed to infection through the consumption of ice that was fabricated with the river water, as the governor was known for absorbing "too much ice."[44] Panic and consternation spread throughout the city, prompting flights of those who could afford it and religious devotion from those who could not escape. But the epidemic subsided not only because many residents had deserted the city but also and mainly because of changing environmental factors in which diarrheal diseases flourish. Indeed, the return of the rainy season contributed to the reduction of exposure to filth—since heavy rains remove raw waste (hence, waterborne pathogens)—and to the reduction of the level of salinity in the water in the two branches of the Senegal River surrounding Saint-Louis. But it should be emphasized that Governor Pinet-Laprade's death to cholera on August 18 put the water supply scheme on hold.

In September, François-Xavier Valière, Navy infantry colonel, was appointed Governor. The source material is far less abundant for the 1869 cholera epidemic mortality and morbidity than for the outbreaks of 1868 and 1893. Sources are silent about the last recorded cases and the ways in which the 1869 cholera epidemic ended. Very little evidence is available concerning the incidence of the disease in the city, but the increasing disease threat had serious political ramifications.

## The Political Impact

The succession of three major epidemics in three years—one yellow fever and two cholera epidemics—which killed or disabled hundreds of city residents, including European civil servants and troops, raised doubts about the expansionist policies under way and the feasibility of empire because of shortage of personnel and the crystallization of the negative image of Senegal as a land of heat and disease. The Minister of Navy and Colonies in Paris questioned the utility of the policy of annexation of peripheral territories not under direct administration and turned to Saint-Louis for specific proposals. Paris was willing to abandon to the British its possessions in the Gambia between the Dembia and Shebar Rivers if "the British government does not impose upon us more sacrifices than that of our suzerainty upon Melacorée and the Moreah lands that depend on it."[45] The fate of other zones of influence, including Toro, Dimar, Walo, Cayor, Saniokhor, and Diander, was also under examination.

These questions were discussed during a special meeting of the *Conseil d'Administration* held on April 11 and 12, 1870. The *Ordonnateur* probably made the most compelling case for the repudiation of the "system of annexations" pursued by Governor Léon Faidherbe and his successor Governor Pinet-Laprade. He argued that the policy of territorial expansion that led to the transformation of Senegal from a vast *comptoir* into a "true colony" in the last twelve years was a costly mistake, a "utopia." He took great pain to explain that the intended goal of creating agricultural or industrial enterprises in these new territories never materialized, that no agricultural centers were created, and that pilot projects for the culture of cotton were even tried and abandoned. Probably his most persuasive argument focused on the fact that the control of the annexed territories was only possible through the means of pressure and unending military expeditions, and that all efforts to assimilate the local populations and to inculcate in them the French culture, mores, and language, in short, to "civilize" and "moralize" them, had failed, at the exception of a few centers where a direct administration was exercised. In his perspective, the new focus would be to maintain a "purely commercial colony." He proposed a new policy centered on four priorities: 1) keeping the trading posts along the Senegal River and along the lower coastal region open; 2) closing the posts in the interior of Cayor, N'Diagne, and Khaulu, at the exception of the posts located on the coastal area between Saint-Louis and Dakar in order to protect the lines of transportation and telegraphic communication, including Gandiole, Betete, M'Bidgem, Pout and Thiès; 3) suppressing the unpopular and inefficient head tax; and 4) allowing the local population to choose their own chiefs as it was the practice before the French intervention that imposed chiefs with little popular support.

The *Ordonnateur* was convinced that the proposed policy changes would not harm the interests of commerce for two reasons: first, the French presence had created among the local population new needs that would force them to sell their commodities and to seek French products in colonial cities and trading posts; and second, led by their legitimate chiefs, the rural communities would unite to repel attacks from any raiders who would invade their lands.[46]

While admitting that the policy of annexations had failed, the Chief of Judiciary Service opposed to return to a status quo ante which existed in regions like Cayor, where the Wolof aristocracies had "oppressed" and "exploited" their subjects prior to French intervention. Instead, he argued in favor of transforming the occupied territories into a protectorate based on treaties, getting to know better these populations, their habits, and their social situation in order to better govern them, and improving their working conditions. In his view, such approach had the advantage of allowing the people under protectorate "to keep their territory, their nationality and their mores," of looking at the French as "benefactors and not as enemies," and of promoting French commercial interests that went hand in hand with "the progress of civilization." At the end of the debate, the *Conseil d'Administration* recommended 1) keeping the territory of Walo because of its economic potentialities and its closeness to Saint-Louis; 2) keeping Diander, given its strategic location as trade route for the caravans heading to Rufisque; 3) the withdrawal from the recently conquered Toro and self-determination for the former subjects of the defeated holy warrior Amadu-Seku; 4) the withdrawal from Dimar while keeping a watchful eye on the Moors who would be tempted to take advantage of the situation; 5) the withdrawal from Saniokhor; 6) and the withdrawal from Cayor where Damel Lat Dior, although recently defeated, still had supporters and could return to power and pose serious threats to French interests in the territory, including the telegraphic line, the telegraphic posts at Gandiole and Betete, the corridor between Niayes and the sea used by mail carriers and travelers. It was agreed that, in the worst case scenario, Saint-Louis could request from the future leader of Cayor to respect the treaty concluded on February 1, 1861 between Governor Faidherbe and Damel Makodu, king of Cayor.[47]

Concerning the future of the trading posts along the Senegal River, the majority of the *Conseil d'administration* members recommended keeping them open for the protection of French merchants and the facilitation of commercial transactions. The *Ordonnateur* even suggested the withdrawal from Medine because of its location so far in the interior but without obvious economic advantages. Other recommendations included keeping the posts of Gandiole and Betete in Cayor and withdrawing from Khaulu and N'Diagne.

Although during the deliberations of the *Conseil d'Administration* no mention was made of the shortage in personnel due to disease threat as a contributing factor in the policy changes, the timing and the reach of the suggested changes suggest a correlation between cause and effect. There was shortage of European personnel as a consequence of the recent yellow fever that killed half of the European population and the two cholera epidemics.

## The Water Supply System

The issue of the establishment of a water supply system was again discussed by the *Conseil d'Administration* during a meeting held on July 12, 1870. The Interior Director, Trédos, indicated in his briefing that the execution of the project would require the mobilization of all the Navy infantry artillerymen, all the *Disciplinaire* personnel, and a section of the *Tirailleurs* battalion. He underlined the fact that the tasks they would perform, that is, digging and filling trenches, laying pipes, and sealing joints, would be detrimental to the health of the Europeans involved in the project. He also pointed out that the project would not cost less than 800,000 francs, that the administration could only mobilize 590,000 francs, and that, given the war between France and Germany, no funding would come from Paris. Governor Valière recognized that his administration had two options: either to postpone the construction of the waterworks or to execute the project for the amount of 685,000 francs. But at the end of the meeting no decision was taken.[48] Nevertheless, the administration sent the Lampsar project to Paris for analysis and approval. In the meantime, floods during *hivernage* seriously had damaged the dam built at the Kassack flood plain, and the reservoir was transformed into a stream. The damage had revealed the vulnerability of the project and raised concern among the decision makers. A new solution had to be found.

The members of the *Conseil des Travaux de la Marine*, an institution based in Paris, met on October 11, 1870 to discuss the viability of the Lampsar project. They recommended that the administration try to build another artesian well in Saint-Louis, to find ways to supply the Kassack reservoir with water-intake pipes far upstream for a longer time period, to provide them with complete information concerning the quality of the water from the reservoir, and to put the pumping station at Makhana.[49] Two years passed without any tangible progress made concerning the construction of the project.

In 1872, the *Capitaine de génie*, Gouin, abandoned the Lampsar project, and designed a new water supply system focusing on the construction of eight vast reservoirs on Pointe Nord in Saint-Louis that could be covered and filled with the river water during *hivernage* through pumps propelled by the northwest winds. Each reservoir would have the capacity of 4,000 $m^3$ and would cost 75,000 francs. Gouin estimated the total cost of the project at 600,000 francs, and the annual operating expenses were estimated at 8,000 francs. According to his calculation, each resident would receive 10 liters of fresh water per day for 20 centimes.[50] But Gouin's project, despite the advantages it had in terms of labor and cost, did not find supporters among the decision makers for reasons that remain obscure. A year later, Governor Valière sent a letter to the Minister of Navy in Paris requesting that an engineer from the service of *Ponts et Chaussées* be sent to Saint-Louis for the establishment of a water supply system at Lampsar. He argued that the cost of maintaining and repairing the damages done to the dam by flood would mount to 8,000 to 10,000 francs per year.[51]

The next five years were spent in political squabbling over cost, technology, and funding. Important segments of the city's residents petitioned for a reliable water supply system.[52] In early 1878, the Navy Ministry in Paris, under the pressure from the *négociants* living in France, rejected the most recent Lampsar project presented by engineer M. Badois because the cost, estimated at 1,505,000 francs, was too prohibitive for the local administration as well as the municipality. But for "political as well as humanitarian reasons," the Governor decided to keep the Ministry's position secret until further notice. Instead, in order to avoid an open hostility between the business community and city residents, and to give moral satisfaction to city dwellers, he requested that the Ministry provide him with a hedge: the submission of the new Lampsar project and the Badois contract annexed to it to the *Conseil des Travaux de la Marine* for examination. Personally, he believed that there were signs that in the near future the business interests would no longer oppose the execution of the project because they would be in the process of demanding the creation of a *Conseil Général* for Senegal with important budgetary prerogatives. He concluded his letter by emphasizing that the execution of the project would constitute "the most important achievement of any administration." The improvement of living conditions would put Senegal on an equal footing with the other colonies.[53]

As part of its internal policy, the administration had to make a complicated decision: to gain time by continuing to give the impression that it was committed to the execution of the Lampsar project. Lacking sufficient financial resources, the governor turned for funding to the municipality which, since 1872, had the authority to borrow money. When the Municipal Council decided to take on debt, a group of 31 prominent taxpayers sent, on April 27, 1878, a letter of protest to Louis Brière de l'Isle, former governor of Senegal and then Minister of Colonies, in which they argued that they were not consulted by the Municipal Council, as required by article 58 of the decree of August 10, 1872. But the Governor had based his initiative on article 42 of the law of July 18, 1837 which stipulated that the Municipal Council had the obligation to consult the *habitants* only when the municipal income was below 100,000 francs.[54]

## Table 3.5. The Revised Estimate of the Badois Contract (in francs)

| LABOR/EQUIPMENT | COST (FRS.) |
| --- | --- |
| Makhana's pumping station | 60,000 |
| Buildings | 78,000 |
| Machines | 180,000 |
| Water transportation | 781,237 |
| Accessory pillars at Makhana | 20,000 |
| Reservoir at Sor | 75,000 |
| Reservoir at Saint-Louis | 75,000 |
| Hydrants/Fountains/Street taps | 55,000 |
| General expenses | 134,000 |
| Total | 1,675,500 |

Source: ANFSOM Sér. géogr. Sénégal XII/18c, Revised estimate (Badois' project), August 28, 1878.

The Governor also spent a great deal of time dealing with the logistics of the project, especially the search for a location for the high-level reservoir and the revision of the estimate. The powder house (*poudrière*) was appropriated to become the water tank, and the initial estimate made by Badois was revised and adjusted. The new estimate of the Lampsar project was 1,675,500 francs.

The estimate did not take into account the expenses related to the distribution pipes through the city, the construction of intermediary reservoirs along the distribution network, and the work to be done at the Gorun flood plain. The strategy adopted by Governor Valière finally paid off. The Ministry of Navy and Colonies in Paris, Vice-Admiral Jauréguiberry, who had served in Senegal as Governor between 1861 and 1863, agreed to back up loans for a water supply system in Saint-Louis, and to put the project out for bids.[55] In December 1879, the contract was given to J. Le Blanc, engineer and supplier, who committed himself to providing Saint-Louis with a reliable water supply system within a reasonable time. The initial work then cost 1,100,000 francs; but additional funds were needed to complete the reservoir in Sor, the reservoir in Saint-Louis, the distribution systems for Sor and Saint-Louis, and miscellaneous expenses. Thus the total cost was estimated at 1,463,000 francs.[56]

The news reached the administration in Saint-Louis on January 20, 1880, and the happy Governor hastened to communicate it to the population: the administration and the Municipal Council were authorized to borrow money for the execution of the Lampsar project.[57] City residents also learned that in order to accomplish the task with efficiency, J. Le Blanc delegated his authority to Mr. Colot, a local engineer.[58] The decision came at the time when the financial cost of operating and maintaining the ship for water distribution had sharply increased for reasons that are not yet clear; it went from 13,368.56 francs for 1,386,251 liters of water distributed in 1872 to 47,571.83 francs for 1,912,051 liters in 1880.[59]

The resolution of the financial question was just one of the many problems with which the administration had to deal. By May 1881 the initial excitement gave way to anxiety. The Lampsar water, measured by Richard, the Navy pharmacist, contained a significant quantity of vegetable debris and other impurities. It did not cook vegetables; beans boiled for two hours with the Lampsar water remained unchanged, while beans cooked with the water from the hospital cistern were well done. In short, the Lampsar water was unfit for human consumption and for most aspects of domestic use.[60] Some members of the *Conseil Général* argued that the engineers of *Ponts et Chaussées* had known all along that the Lampsar water was not even close to safety standards, but, for selfish or other unknown reasons, they misled the administration to support a project bound to fail.[61] These reactions underscored the tension between the French, who controlled the colonial administration, and the *métis*, who dominated the *Conseil Général* as well as the Municipal Council, and who stood for the protection of local interests.

The administration continued to borrow money for Saint-Louis' water supply system, despite the succession of accusations and counter-accusations. In 1886 the project was completed. The water was pumped out of Makhana between December and June and out of Khor between July and early December. But when it arrived, Saint-Louis' residents realized that the water distribution network was incomplete. Certain sections of the city either were deprived of water for long hours during the day or received only a small quantity of water. Thus, 12,000 additional meters of smaller distribution pipes and more street taps and hydrants were needed. The total cost was estimated at 80,000 francs. The peri-urban slums were provided with a few public fountains.[62] But the extension of the water distribution network through the city did not take place overnight. It was accomplished over the course of years and in several stages.

The water itself, between January and April, was pure and contained a small quantity of organic matter, salt, and clay that could be eliminated through a public sand-filtration system. But toward the end of the dry season (May–June), samples of the water from four public taps in the city and in Sor revealed that the water had become brackish, not drinkable, and loaded with vegetable matter called "*tambalayes*."[63]

The population growth that characterized the 1890s increased the demand for water, resulting in a deficit of approximately 800 tons. The shortage, mostly toward the end of *hivernage* (early November), obliged the administration to tap the small arm of the Senegal River to supply the military hospital and the Roignat garrison

Table 3.6. The Cost of Bathing at the Military Hospital, Saint-Louis, 1892 (in francs)

| SERVICE | COST | BENEFICIARIES |
|---|---|---|
| Partial bath | 1.50 | Officers |
| | 2.00 | Private individuals |
| Body | 2.00 | Officers |
| | 3.00 | Private individuals |
| Shower | 1.50 | Officers |
| | 2.00 | Private individuals |

Source: ANS, Bulletin Administratif, 1892, 123–24.

and the big arm of the river to meet the needs of the Subsistances Service; and, later on, officials used the "*Akba*" ship to bring water from upstream near Richard Toll to the city. By then physicians became alarmed by the trends in incidence of all types of bowel infections, labeled as "dysentery," among city residents as a result of access to polluted river water. Cases of "dysentery" were the fourth-largest cause of hospital admissions in 1889, the third in 1891, and the second in 1893 and 1895. Dysentery was followed by typhoid fever, which dominated the category of "fevers."[64] The working class and the urban poor paid the heaviest toll to water-borne diseases.

The quantities of water were insufficient to cover government employees' daily needs. Bathing in particular was a crucial problem. The bath was available at the military hospital between 2 and 4 p.m. every day on a fee basis for those who carried an authorization from the head of the Administration Service and a medical certificate.

The table indicates that during periods of water shortage, only the privileged elite could afford the cost of bathing. When cholera struck Saint-Louis in June–October 1893, there was a serious water shortage in the city.

## The 1893 Cholera Epidemic

The wide gap between the two cholera epidemics (1869 and 1893) deserves an explanation before focusing on the new epidemic. Since the environmental factors that provide the breeding ground for cholera—including poor sanitation, lack of hygiene and crowding living conditions—were still present in the city, how can we account for the relative absence of the cholera epidemics? There are three main reasons. First, it is possible that the aquatic environment of Saint-Louis and the variation in the water temperature and salinity did not allow *V. cholera* to survive for a long time on fish or shellfish, to become endemic, and to contribute to seasonal epidemics. Second, the policy of forced removals of the urban poor from the city center had helped reduce the exposure to raw waste, hence, waterborne pathogens. Third, and finally, the disinfection done in the framework of anti-yellow fever measures and later the use of filtration prevented the spread of cholera. But by the 1890s Saint-Louis had grown in size, which contributed to its vulnerability to a threat of pandemicity.

The fifth cholera pandemic started in India in 1881 and spread to Mecca and the Mediterranean region two years later before spreading to the rest of the world by land and sea. Beginning in June 1890 Paris kept the administration in Saint-Louis informed about the presence of cholera and yellow fever in Spain. In order to prevent the disease from entering and spreading in Senegal, a quarantine of sixteen days' observation was imposed against all ships arriving from Europe and the Canaries Islands in accordance with the decree of August 29, 1884. Merchandise, parcel post and letters were disinfected, and the import and sale of fruits and vegetables was prohibited.[65] In April 1892, cholera re-emerged during the Hartwar festival following the dispersal of the Hindu pilgrims and continued its progression by land and sea. In Saint-Louis, the quarantine remained in effect until October 27, 1892, when it was lifted for all ships but those arriving from Hamburg where cholera was raging since August and had produced 8,594 deaths by mid-September.[66] Once alarmed by the disease threat, the 1892 International Venice Conference put in place a surveillance system to combat cholera, established a sanitary quarantine station and a lazaretto at the Suez Canal, and organized the El Tor Station for the surveillance of the Persian Gulf. In March 1893, the Paris International Sanitary Conference (for the third time) focused on the protection of Europe against cholera through the regulation of Muslim pilgrims, the organization of a quarantine station in the Gulf Region, the supply of medication to Hejaz during Ramadan, and the creation of a "sanitary police."[67]

During the *hivernage*, on June 29, a ship dropped anchor at Saint-Louis' harbor and started unloading the commodities. One of the daily workers who had helped carry the merchandise went back home in the northern part of Saint-Louis and became ill. He complained of severe cramps, vomiting and diarrhea. When Dr. Charles Carpot, a métis from one of the most powerful families of Saint-Louis, visited him in late afternoon, he found him dead. His diagnosis, probably through clinical observation, was clear: it was cholera. Two days later, during the night of July 1 and 2, a woman in good health fell sick after eating dinner, and died early in the morning following long hours of incredible suffering.[68]

Dr. Carpot reported the cases to the Mayor and the Head of Health Service. On July 2 and 6, the Hygiene Council met to discuss the necessary measures to be taken in order to contain the spread of the disease. By then the number of deaths rose to an average of eight per day. New cases were reported in different parts of the city. Mayor Jules Couchard signed an ordinance on July 6 containing a comprehensive anti-cholera program. He urged the landlords, tenants, and keepers or guardians of houses and huts to report to him new cases of cholera as well as deaths. City residents were also ordered to put biochlorure of mercury in vases containing the patients' vomiting and stool, to bury the dead rapidly, even during the night, and to burn the mattresses and other objects used by cholera victims. Article 11 of the ordinance required all the houses and huts occupied by the *indigènes* in the northern *quartier* of the city-island from Boufflers Street upward and in the southern *quartier* from Repentigny Street southward to be submitted to an automatic and general disinfection.[69]

The underlying assumption behind the Mayor's ordinance was that cholera was the "Black man's disease" and that the urban poor were the natural victims of cholera not only because of their ignorance of hygiene and poor living standards, but also because of their culture, especially the burial customs and the imperatives of solidarity. According to Dr. Charles Carpot, one of the principal causes of the sudden and simultaneous outbreak of the epidemic in different *quartiers* had to do with

> the complete ignorance of the notion of hygiene and sanitation on the part of the residents of the outlying *quartiers* (...and with) the requirements of the customs of the land that when a person died all the acquaintances had to visit the parents of the deceased in order to offer their sympathy.[70]

Dr. Charles Carpot correctly mentioned that the gathering of many people at the burial site and at the home of the deceased contributed to the spread of cholera among the urban poor, who formed the majority of the victims because they had no access to sanitary toilets. But he lost sight of the fact that the urban poor's access to clean water was limited, that they continued to use the contaminated river water to wash vegetables and clothes, rinse the containers used to transport kitchen waste to the river, and, occasionally, for domestic consumption.[71] Moreover, he failed to integrate Dr. Koch's findings in his explanations, chief among them being the idea that without the import of a specific microorganism, epidemic cholera would not arise, whatever the predisposing factors.[72] For cholera was not only the product of human agency, of social inequality, but it was also a disease of empire, as R. Evans pointed out.[73] Dr. Carpot's attitude was consistent with the scapegoating mentality discussed above.

In any case, the police were responsible for the follow-up of the implementation of the provisions of the Mayor's ordinance. The daily report made by the Police Chief included the number of deaths per day as well as the number of persons who contravened the ordinance by throwing trash in the street or by failing to sweep in front of their houses.[74]

The re-emergence of cholera created panic among the *indigènes*, the majority of whom refused to be treated at the existing medical facilities. They nevertheless welcomed the measures of disinfection of their houses and the incineration of the ob-

112 | COLONIAL PATHOLOGIES, ENVIRONMENT, AND WESTERN MEDICINE

jects used by the victims. The available health data do not help explain their refusal to get medical attention during this terrible epidemic crisis, but resistance could result from the cultural insensitivity and cross-cultural mis-communication exhibited by the health authorities.

It was not until the following day, July 7, when the Sanitary Commission officially declared the re-emergence of cholera in the city. The number of deaths had

Table 3.7. Daily Reported Cholera Deaths, Saint-Louis, 1893

| DATE | DEATHS | DATE | DEATHS | DATE | DEATHS |
| --- | --- | --- | --- | --- | --- |
| 1 July | 2 | 23 | 38 | 14 | 2 |
| 2 | 4 | 24 | 24 | 15 | 1 |
| 3 | 5 | 25 | 35 | 16 | 1 |
| 4 | 9 | 26 | 41 | 19 | 1 |
| 5 | 7 | 27 | 43 | 20 | 1 |
| 6 | 11 | 28 | 32 | 22 | 2 |
| 7 | 8 | 29 | 23 | 26 | 1 |
| 8 | 20 | 30 | 23 | 4 Sept. | 1 |
| 9 | 16 | 31 | 33 | 5 | 2 |
| 10 | 18 | 1 Aug. | 20 | 6 | 1 |
| 11 | 16 | 02 | 15 | 7 | 11 |
| 12 | 26 | 03 | 19 | 11 | 1 |
| 13 | 22 | 4 | 10 | 13 | 4 |
| 14 | 28 | 5 | 6 | 15 | 1 |
| 15 | 16 | 6 | 9 | 18 | 2 |
| 16 | 23 | 7 | 6 | 19 | 4 |
| 17 | 30 | 8 | 6 | 21 | 1 |
| 18 | 28 | 9 | 5 | 2 Oct. | 2 |
| 19 | 25 | 10 | 2 | 7 | 1 |
| 20 | 38 | 11 | 8 | 11 | 1 |
| 21 | 56 | 12 | 3 | | |
| 22 | 53 | 13 | 1 | | |
| Total: 897 deaths | | | | | |

Source: ANS AOF H 36.2, Rapport journalier du Commissariat de Police de Saint-Louis.

reached an average of 20 per day on July 8, and 56 deaths were reported on July 21 alone, as the following table indicates.

By the end of July, Saint-Louis had suffered 753 deaths among the urban poor, out of a population of 20,173.[75] Only four Europeans had died from cholera. In early August, the case rate started to decline with fewer than ten cases per day. The last case was reported on October 11. By then cholera had killed 897 poor city residents, according to official records. But missionaries reported between 1,500 and 2,000 deaths.[76]

The missionaries interpreted the ravages of cholera in religious rather than class terms. They believed that the Christian community had received a special protection from "the Providence." When cholera appeared in Saint-Louis, parishioners turned to God, organizing processions and prayer meetings in Sor where the statue of Saint Roch was erected. Christians were convinced that the Virgin Mary had provided them with protection in reward for the cathedral recently built in Sor as a proof of their trust and devotion. Missionaries' reports indicated that

> Mary rewarded our confidence, and we escaped the disaster. The *marabouts*, taken by surprise, expressed their amazement for the fact that the Christians did not die, while their own devotees died at the rate of 50 to 60 per day.[77]

## Figure 3.2. Daily Reported Cholera Deaths, Saint-Louis, 1893

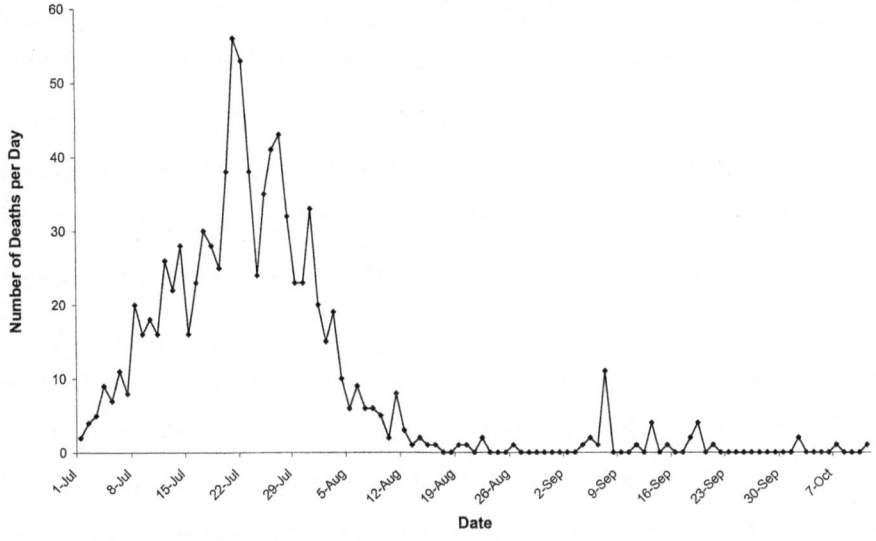

Source: ANS/AOF/H/36.2, Rapport journalier du Commissariat de Police de Saint-Louis.

The missionaries' assessment of the incidence of cholera in the city needs to be understood in the context of the ongoing competition between the missionaries and the *marabouts* for bodies and souls. Missionaries saw the selective nature of cholera as a message informing the *marabouts* about the Christians' strength in divine resources. They made statements in words as well as in actions (processions, prayers) about their command over supernatural resources. Their objective was to destroy the *marabouts*' reputation, and to show the Muslims that the Christians were on the winning side, and that conversion to Christianity was the right move. Thus, from the missionaries' perspective, the "God of Christians was stronger than the God of Muslims."

## Cholera Aftermath

The re-emergence of water-borne diseases, especially cholera, remained one of the main preoccupations of the administration, given that by the 1890s Saint-Louis had grown in size. Alarmed by the news of a fresh cholera outbreak in Mecca in 1895, the European health authorities agreed to extend the quarantine period from five to ten days and to enforce municipal cleaning. In Saint-Louis the quality of the river water taken in November 1895 in front of the military hospital and the Roignat garrison revealed that Saint-Louis' water continued to contain organic matter and other pathogens (*E. Coli* and *Bacillus Typhus*) responsible for diarrheas, dysentery, and typhoid fever. In order to reduce the high incidence of the diseases of the digestive system, the Chief Medical Officer recommended the division of each cistern into two compartments: the first would contain the water for human consumption from the rain collected during *hivernage* and the "*Akba*" water supplied during the dry season; and the second compartment would contain river water that could be used for cleaning, bathing, washing, and, if boiled, cooking and *tisane*-making.[78] But the *Commandant en Chef des Troupes* made a different suggestion to the governor general: to equip each military settlement with two separate cisterns, one cistern containing the water to be used for human consumption, and the other containing the water for other domestic uses (cleaning, bathing, cooking, etc.). He recommended that, following the wishes expressed by General Borgnis-Desbordes during his visit to Senegal in 1892, the water be filtered (using Chamberland filter type) as it was a routine practice in the metropolitan army. He reminded the governor general that the issue of improving the quality of water through filtration had been brought to the attention of the Ministry of Colonies in 1894, and the Undersecretary of State for Colonies in early 1895, but it had not been resolved because of the heavy operating expenses related to the use of filters. He concluded that if the administration was determined to preserve the colonial society and the very existence of the European soldiers, the officials needed to obtain special loans in order to buy the much-needed equipment.[79] A temporary solution was found in 1896 when the authorities decided to stop taking water from the Senegal River around Saint-Louis, which was polluted and contained pathogens such as *E. coli*. Instead, the city would be supplied with the water from the rain stored in the cisterns during *hivernage*, and with the river water from upstream distributed during the dry season.[80]

Thus, almost ten years after the establishment of the water supply system in the city, the demand for clean water was far from being met for natural, financial, and technological reasons. The water from the Kassack flood plain, taken from a pumping station at Makhana since December, was polluted by March: it was loaded with vegetable debris and animal waste, contained a high level of salt, and had a bad smell. It was not even good for sprinkling gardens. By May, the flood plain was dry, forcing the pumping engines to shut down for many hours every day. The water technology used was not well developed in those years. The level of salt in the water often damaged the pipe transporting the water from Makhana, thus reducing the volume of water transported from 2,200 to 1,800 m$^3$ per day.[81] Alternative projects were tried at Richard-Toll and N'dial and abandoned.

In order to find a permanent solution to the problem of deteriorating water quality, the *Conseil Général* approved, in May 1899, an administration plan aimed at improving the water supply system, and authorized a 700,000 francs loan for the building of a new dam at Kilère (150,000 francs), the repair of the main distribution pipe near Makhana (80,000 francs), the installation of sand filters at Khor (80,000 francs), the extension of the distribution pipes network in the city, especially in the Pointe du Nord *quartiers*, as well as in the peri-urban villages of Guet-Ndar and Sor (80,000 francs), and miscellaneous tasks (55,000 francs).[82] In a report sent to the *Conseil Général*, the Head of the *Travaux Publics* Service, Malenfant, suggested to improve the quality of the Makhana water by purification before its distribution. Such an operation required either the construction of covered reservoirs measuring 40,000 m$^3$ and costing 1,800,000 francs, or the recourse to a less expensive operation of "*ozonization*" used at Lille and Osborne, and costing 350,000 francs.[83] But the implementation of the new plan was delayed by a terrible yellow fever epidemic that struck the city in 1900.

## Conclusion

The emergence of lethal epidemics in the 1860s seriously impacted the expansionist policies during the formative years of French empire in West Africa. The two cholera epidemics that struck Saint-Louis revealed the difficult living conditions in the colonial city, the dimensions and persistence of racial inequality and racialized policies in the delivery of urban services, especially fresh water supplies during the dry season. The management of cholera provided a window into the processes of urban city planning and city improvement. In particular, efforts to make the city healthier by installing the water and sewer systems helped understand the operations of the colonial bureaucracy and the decision-making process between the central administration in Paris and the local administration in Saint-Louis concerning the funding of the colonial urban infrastructure. The involvement of Paris in financing the colonial urban infrastructure shows the gap between the colonial doctrine of financial self-sufficiency for each colony and the practice. Although separated by twenty years of interval, the two cholera epidemics highlighted the continuity of the image of tropicality of Senegal as well as the perception that cholera was the "Black man's disease" despite the progress made in the understanding of disease etiology. The anxiety about the tropics would be heightened once again with the outbreak with another yellow fever epidemic, as the next chapter will show.

## Notes

1. Paul Farmer, *Infections and Inequalities: the Modern Plagues* (Berkeley: U. of California Press, 1999), 4.
2. Gilbert-Pierre Eyoum, "Le choléra en Afrique Noire Francophone," Doctoral dissertation in medicine, Saint Antoine, Paris (n.d.), 14–15.
3. Ibid., 16.
4. Rita R. Colwell, "Global Climate and Infectious Disease: The Cholera Paradigm," *Science* 274 (Dec. 20, 1996), 2026.
5. Gilbert-Pierre Eyoum, "Le choléra en Afrique Noire Francophone," 23.
6. Ibid., 25.
7. Rita R. Colwell, "Global Climate and Infectious Disease," 2026.
8. Ibid., 2025.
9. F. M. Snowden, *Naples in the Time of Cholera, 1884–1911* (New York: Cambridge University Press, 1995), 16.
10. R. J. Evans, *Death in Hamburg : Society and Politics in the Cholera Years, 1830–1910* (Oxford: Clarendon Press, 1987), 227.
11. David Arnold, *Colonizing the Body. State Medicine and Epidemic Disease* (Berkeley: U. of California Press, 19936) 194.
12. Richard J. Evans, *Death in Hamburg*, 268–69.
13. David Arnold, *Colonizing the Body*, 194.
14. Charles L. Briggs and C. Mantini-Briggs, *Stories in the Time of Cholera. Racial Profiling during a Medical Nightmare* (Berkeley: University of California Press, 2003), 1.
15. Christine Matthieu, "Le cholera a Otavalo (Equateur): Une épidemie qui revèle les ruptures socio-culturelles," Doctoral thesis in medicine, Université de Nancy I, 1992, 141.
16. F. M. Snowden, *Naples in the Time of Cholera, 1884–1911*, 22.
17. Christine Matthieu, "Le cholera a Otavalo (Equateur)," 136.

18. Christine Matthieu, "Le cholera a Otavalo (Equateur)," 138.
19. Snow, "Snow on Cholera (New York, 1936)," in F. M. Snowden, *Naples in the Time of Cholera*, 23.
20. ANS/H48-1, Mission Sanitaire du Sénégal (Février–Mars 1901): Rapport Technique, by engineer Jacquerez.
21. Sometimes, the river water remained salty until August. See Moniteur du Sénégal, August 2, 1868, 218; see also ANFSOM. Séries géographique Sénégal XII d. 17 b, "Mémoire sur un avant-project de conduite d'eau destinée à l'alimentation de Saint-Louis et de ses faubourgs," April 6, 1869 by the chief of Bureau du Génie.
22. Ministère de la Marine et des Colonies. *Sénégal et Dépendances, Bulletin Administratif des Actes du Gouvernement, 1870* (Saint-Louis: Imprimerie du gouvernement, MDCCCLXXI), 8.
23. *Bulletin Administratif*, 1892, 119–20.
24. ACSE, Chevilly-Larue, Boîte no. 157 B. Sénégambie. Sénégal. Gambie. Affaires diverses, 1864–1876, piece 033: "Epidémie du Sénégal," in *Univers*, Nov. 23, 1867.
25. ANS, *Moniteur du Sénégal et Dépendances*, Jan. 26, 1864, 14.
26. ANS/H2/AOF/9, médecin 1st class, Beaussier, to médecin en chef, Dr. Julien-Henri Rulland, Jan. 29, 1868.
27. ANFSOM, Série géographique Sénégal d. 18 a, minutes of the Conseil d'Administration meeting, Febr. 4–5,1878.
28. ANS, *Bulletin Administratif*, 1885, 534; see also the decree of March 21, 1885.
29. ANS/H20, Séries Sénégal, minutes of the HPSC meeting, Jan. 10, 1868.
30. ANS/H2/AOF/38, médecin en chef to governor, n.d.
31. ANFSOM, Séries géogr. Sénégal XII 17 b, "Rapport sur le forage du puits artésien de Saint-Louis," by the Inspecteur du Service des Ponts et Chaussées.
32. ANFSOM, Séries géogr., Sénégal XII d. 17 b.
33. ANFSOM, Séries Sénégal XII 17 b, "Mémoire complémentaire sur le projet de conduite d'eau du marigot de Lampsar à Saint-Louis," by the director of Ponts et Chaussées, May 12, 1870.
34. ANFSOM/Séries géogr. Sénégal XI/30, interim Chief Medical Officer, Henri Rulland, to Governor: "Rapport sur l'épidémie de choléra de Saint-Louis (Sénégal) en 1868," January 3, 1869, 1–5. See also ANS/H27 no piece nb.
35. ANFSOM/Séries géogr. Sénégal XI/30, Henri Rulland to Governor, Jan. 3, 1869, 11.
36. ANFSOM/Séries géographique Sénégal XI/30, Henri Rulland to Governor: "Rapport sur l'épidémie de choléra de Saint-Louis," Jan. 3, 1869. In his doctoral thesis in medicine, Gilbert-Pierre Eyoum mistakenly presents these mortality figures for Dakar instead of Saint-Louis; see Eyoum, "Le choléra en Afrique Noire francophone," doctoral thesis in medicine, Université P. et M. Curie, Paris VI, 1986, 22.
37. ACSE, Bulletin Général 6 (June 1868–March 1869), 862–64; see also Sister Marie-Ange, 79.
38. ANF/CAOM/2B47, Governor Jauréguibery to minister, June 7, 1877 related to the electoral lists and movements of the population.
39. ANS/H27, Dr. Henri Rulland to Governor: "Rapport sur l'épidémie de choléra de Saint-Louis (Sénégal) en 1868," Nov. 25, 1868, 5.
40. Dr. Rulland's thesis about the Moroccan origin of cholera was later rejected by Dr. A. Thiroux, who favored the "local pollution" thesis, arguing that the V. cholerae was already present

in the river water around Saint-Louis as early as 1866 based on the descriptions made by Vauvray. For details, see A. Thiroux, "Le N'diank, choléra du Sénégal; son agent pathogène," *Bulletin de la Société de Pathologie Exotique*, 5 (1912), 753–62.

41. ACSE, Bulletin Général 6 (June 1868–March 1869), 864.
42. ANFSOM/Séries géogr. Sénégal XII/17b, "Mémoire sur un avant-projet de conduite d'eau destinée à l'alimentation de Saint-Louis et de ses faubourgs," by the head of Bureau du Génie, April 6, 1869.
43. ACSE, Annales, 23–25; see also Bulletin de la Congrégation 7 (1869–70), 131–2, 135–7.
44. A. Thiroux, "Le n'diank, choléra du Sénégal," 760.
45. ANFSOM/FM/SEN/VII/22C, minutes of the Conseil d'Administration meeting, Apr. 11–12, 1870.
46. Idem.
47. Idem.
48. ANFSOM, Séries géogr. Sénégal XII/17b, minutes of the Conseil d'Administration meeting, July 12, 1870.
49. ANFSOM/Séries géogr. Sénégal XII/17b, minutes of the Conseil des Travaux de la Marine meeting, Oct. 11, 1870.
50. ANFSOM/Séries géogr. Sénégal XII/17c/14, "Extrait d'un projet de réservoir à la pointe du nord de l'île de Saint-Louis," by the Capitaine de génie, Gouin, s.d. 1872.
51. ANFSOM/Séries géogr. Sénégal XII/17b, governor to minister, Jan. 9, 1873.
52. ANFSOM/Série géogr. Sénégal XII d. 18, governor to ministry, April 18, 1878.
53. ANFSOM/Série géogr. Sénégal XII/18a, governor to minister, April 22, 1878.
54. ANFSOM/Série géogr. Sénégal XII d. 18 a, Habitants to governor, 27 Apr. 1878; 3 May 1878; governor to Habitants, 3 May 1878; minutes of the Conseil des Travaux de la Marine meeting, 4 June 1878.
55. ANFSOM/Série géogr. Sénégal XII/18c, minutes of the Conseil d'Administration meeting, Jan. 2, 1879.
56. ANFSOM/Série géogr Sénégal XII/19, Cahier des charges relatif à l'entreprise des Travaux à exécuter pour l'etablissement d'une conduite d'eau potable à Saint-Louis, Paris, Dec. 10, 1879.
57. The decree of May 29, 1880 authorized the administration to borrow 800,000 francs for the water supply system at the interest rate of 5%; the debt would be reimbursed in 12 years, using the ordinary resources of the colony, especially the taxes on alcohol and customs service. See Moniteur, June 22, 1880, 116.
58. ANS, Moniteur, Febr. 3, 1880, 15–16.
59. ANS, Moniteur du Sénégal et Dépendances, 1873, 1; Moniteur, 1881, 101.
60. ANS, Moniteur, June 14, 1881, 133.
61. ANS, minutes of the Conseil Général meeting, Febr. 6, 1882.
62. ANS, Conseil Général, Récueil des rapports, 1889, 23 and 25. See also Bulletin Administratif, 1889, 26–27.
63. ANS, Journal Officiel du Sénégal et Dépendances, March 28, 1889, 120; August 8, 1889, 291, 301; see also ANFSOM/Série géogr. Sénégal XII/21, Rapport du chef de Service des Travaux Publiques, July 30, 1889.
64. ANS/H16, Tableau synoptique des maladies et mouvement des malades, 1889–1913.
65. ANS/H4/Senegal, minutes of PHSC meeting, June 20, 1890.

66. ANS/H4/Senegal, minutes of the Sanitary Commission, Saint-Louis, Oct. 27, 1892; for daily reported cholera cases and deaths in Hamburg, see R. J. Evans, *Death in Hamburg*, 295.
67. Nermin Ersoy, Yuksel Gungor, and Aslihan Akpinar, "International Sanitary Conferences from the Ottoman Perspective (1851–1938)," *Hygiea Internationalis: An Interdisciplinary Journal for the History of Public Health*, 1.10 (2011), 66–67.
68. ANS/H16, Dr. Carpot, "Nosologie...," 238.
69. ANS, Journal Officiel du Sénégal, July 8, 1893, 30–32.
70. Dr. Ch. Carpot, "Nosologie...," 238.
71. See A. Thiroux, "Le n'diank, cholera du Sénégal," 758.
72. For more on Koch and the 1890s, see Peter Baldwin, *Contagion and the State in Europe, 1830–1930* (New York: Cambridge University Press, 1999), 164–67.
73. R. Evans, *Death in Hamburg*, 474, 564.
74. ANS/H36/2, rapports journaliers du Commissaire de police de Saint-Louis.
75. ANS/H16, Dr. Carpot, "Nosologie...," 238.
76. ANS, Bulletin de la Congrégation, XVII, 205.
77. ANS, Bulletin de la Congrégation, 1892–3, XVII, 205.
78. ANS/H15/2, Général de Brigade Boilève to governor general, Nov. 15, 1895; ANS/H15/3, Médecin-major Henry Reboul to Commandant en Chef des Troupes de l'A.O.F., Nov. 12, 1895; ANS/H15/5, Médecin en Chef to governor general, no. 294 of Nov. 24, 1894.
79. ANS/H15/6, Commandant en Chef des Troupes to governor general, no. 551 of Dec. 13, 1895.
80. ANS/2G/1,7, Rapports des chefs d'administration, 1896: Médecin en chef to governor general, July 15, 1896.
81. ANS, Conseil Général, Récueil des Rapports, minutes of the meeting in May 1899, 26.
82. ANS, Conseil Général, Récueil des Rapports, minutes of the meeting in May 1899, 30.
83. ANS, Conseil Général, Récueil des Rapports, 33–34.

CHAPTER FOUR

# A Conflict of Interests among Commerce, Competing Conceptions of Public Health, and Civil Liberties, 1882–1901

The ravages caused by the great yellow fever epidemics, combined with the injunctions from the International Sanitary Conferences, led the authorities to rethink the organization of the sanitary regime in Senegal and French West Africa (see chapter 3). But the epidemic of fear and suspicion and the epidemic of stigmatization that accompanied yellow fever epidemics pushed for the adoption of the contagionist strategy that relied mainly on strict sanitary cordons and quarantines. Convinced that yellow fever was brought to Senegal by sea, the French medical establishment in Senegal identified the coastal region of West Africa, from south of the Pointe de Sangomar—an island located at the mouth of the Saloum River in the Fatick region of Senegal—to French Congo as the 'foyer of infection,' and prompted the French government in Senegal to act more forcefully by imposing an automatic annual quarantine of ships arriving from regions suspected of contamination. The incriminated space included the following: British Gambia, Portuguese Guinea Bissau, French Guinée, British Sierra Leone, independent Liberia, French Côte d'Ivoire, British Gold Coast, German Togo, French Dahomey, British Nigeria, German Cameroun, Spanish Equatorial Guinea, French Gabon, and Congo. This chapter examines the administrative problems and the conflict of interests that the routinization of quarantines generated not only among various constituencies in Senegal and France but also among Senegal and neighboring colonies and beyond. It also analyzes other drastic steps taken to prevent another outbreak of yellow fever as well as the problems that changing knowledge about disease etiology posed to the medical and public authorities. The main argument here is that the ambitious program designed to contain yellow fever failed because it was based not only on an imperfect science but also on the assumption that the public health structures were adequate, that the past experience with epidemics had generated a public health culture, and that the merchants, ship captains, and the public at large would cooperate with public health officials in charge of the implementation of the new sanitary legislation.

In the wake of the 1878 yellow fever epidemic, the Navy Health Superior Council in Paris defined the problem to be resolved in terms of flaws in the surveillance system, and they urged that the law of March 3, 1822 and the regulations of the maritime sanitary police of February 22, 1876, so far applied in French ports on the Mediterranean Sea against ships arriving from yellow fever–infected lands like Brazil, be implemented 'without attenuations' and 'rigorously in a colony, like Senegal, so predisposed, by its climatic conditions, to receive and develop this disease.'[1] They were convinced that bureaucrats in Senegal were not sufficiently acquainted with these laws and rightly feared that any negligence to implement these rigorous sanitary measures could result in the re-emergence of yellow fever, especially during the upcoming rainy season. They suggested that copies of these laws and regulations be made available to all civil servants.[2] This directive explicitly pointed to the climate of Senegal as the 'predisposing' factor in the outbreak of yellow fever.

The 'sanitation syndrome,' doubled with the medical profiling of the *indigènes*, also inspired the initiatives undertaken since 1882 by both the colonial administration and the city government in Saint-Louis. These initiatives focused on the *Petite Voirie*, the *Grande Voirie*, and the external contacts. The *Petite Voirie* emerged from the municipal budget and managed the garbage collection in the city streets, marketplaces, and other terrains.[3] The *Grande Voirie* was funded by the local (colonial) budget, was executed by the *Service des Ponts et Chaussées*, and focused on the public places, garrisons, bridges, docks, and routes. Measures aimed at protecting Senegal against contagion from external contacts through port-cities consumed the energies of the authorities and revealed very costly in terms of financial and human resources. The drastic measures, inspired by both localist/sanitarian and contagionist approaches, were believed to prevent another epidemic.

## Localists and Urban Hygiene

Fear and panic prompted city officials to review various provisions of public health laws that they judged outdated or inadequate for the current health crisis. A new ordinance, issued by Mayor A. de Bourmeister and dated January 18, 1882, required city residents to clean the confines of their property daily, including sidewalks and sections of the streets in front of their houses and huts, and to rid their property of garbage, stagnant waters, and kitchen waste. They had to throw garbage in the river either before 8 a.m. or after 6 p.m. and kitchen waste any time during the day (art. 1). But the ordinance went beyond concerns for cleanliness to include the urban poor's survival economic activities on the list of 'exciting causes' of the disease and subjected them to the mayor's authorization, including drying animal skins and fish[4] on the city-island, in the interior courts, on the streets, on the roofs of huts, and on tops of walls, slaughtering animals in places other than slaughterhouses (art. 2), and raising pork in the city (art. 4). Women were not allowed to pound millet at night between 10 p.m. and 4 a.m. (art. 10). All the noise associated with cloth making was prohibited between 8 p.m. and 4 a.m. (art. 15).

Aspects of the municipal ordinance prohibiting 'noise' were actually a direct assault on social and cultural norms and religious practices of the urban poor, such as begging in the city streets or at people's homes (art. 8), beating drums, clapping hands, singing on the streets, social gatherings after 8 p.m. without the express authorization of the mayor, and riding horses in the streets. Even the blind, well known for earning a living by singing praises to passerby or in people's homes, were prohibited from 'making noises' (*proférer des cris*). Also mentioned was firing in the air or using fireworks without first securing the mayor's approval (art. 12)

Any initiatives likely 'to obstruct free passage or to reduce safety or freedom' on the streets and sidewalks were no longer tolerated. These comprised activities, such as standing at the street corners or on the sidewalks; stopping camels, horses or oxen used for transportation while riding on the city streets; and unloading merchandise or other objects on the sidewalks (art. 9). In addition, letting domestic animals run loose in the city streets resulted in the payment of fees by the owners (art. 11). The Geôle Bridge was declared off limit for animals; and horses were not allowed on the street leading from the Guet-Ndar Bridge to the beach (art. 13).

The very bodies of the urban poor came under increasing scrutiny; they had to be disciplined. Article 3 specifically prohibited defecation or urination on public places or on the banks of the Senegal River, as well as 'showing, either by bathing, or by cleaning private parts, naked bodies in a way that violates decency.' Here again, the colonial medical discourse deployed the theme of the 'naked bodies' of the natives as representing the 'remnants of savagery.'

The police commissar, who worked under the authority of the mayor on matters of municipal police, was responsible for the daily general inspection of the city-island and weekly general inspection of the commune. He was required to present a weekly written report on the sanitary situation of the city and slums to the mayor and transmit a copy to the interior service chief within 48 hours (art. 5).

Only two provisions of the municipal ordinance specifically targeted Saint-Louis' elite and soldiers; one prohibited the possession of more than eleven kilograms of powder and stocks of petroleum, and the other required owners of cafés, cabarets, and public billiards to stop serving soldiers and navy personnel after 8 p.m. without a written permission from their chefs de corps, not to serve or sell them hard liquors, and to close their properties at 10 p.m., unless they obtained the mayor's permission. The interdiction to sell hard liquors to the military personnel also concerned any merchants who had license to sell alcoholic beverages (art. 16). The provision related to hard liquors actually repeated the restrictions imposed by the ordinance of July 28, 1853, which implemented the ministerial dispatch of 9 May of the same year regulating cafés and cabarets. Violators of the ordinance were liable for a misdemeanor according to the dispositions of the penal code (art. 18). Various provisions of the ordinance were immediately enforceable by municipal agent-voyeur, the police chief, and the chiefs of the slums of Bouetville (in Sor), Guet-Ndar and Ndar-Toute (art. 19).[5]

More evidence is needed to evaluate the ways in which the municipal ordinance was enforced, and whether anybody was ever prosecuted. Suffice it here to underline the fact that the mayor's ordinance reflected the divide, inspired by the epidemiological theory, between sanitary citizens who were supposed to obey the laws, and the unsanitary subjects whose risky behaviors and lifestyle, survival economic activities—such as fishing industry, textile industry, food making and distribution, and transportation—and social practices were perceived as contributing factors to the spread of epidemic infectious diseases and a threat to the city's health. Therefore, city residents, especially the urban poor, had to give up their civil liberties in the name of public health.

As *hivernage* approached, troops were dispatched in 'dissemination camps' in early May 1882 in order to save their lives in case of health emergency. These camps were located at Pointe-aux-Chameaux, Bop-Diara, Gandiole, and in Saint-Louis. Efforts were undertaken to make life comfortable inside the camps starting with daily food supplies and mail delivery.[6] But May and June went by without an outbreak of yellow fever epidemic. The colonial and municipal authorities in Saint-Louis were able to celebrate with great pomp the Fourteenth of July holiday, and to distribute prizes to the winners of various sporting events as well as gifts to the indigents in the hope that 'the memory of the national holiday would remain engraved in their heart as a day of rejoicing.'[7]

Given the emphasis in epidemiological theory on the influence of crowded living conditions on urban mortality, the public health authorities strongly recommended the implementation in the communes of Saint-Louis, Gorée-Dakar, and Rufisque, of the laws of 19 January, 7 March and 13 April 1850 passed in France against unsanitary dwellings, which was accomplished with the issuance of the decree of September 15, 1882. The decree permitted the governor to determine the number of members of the municipal councils who would take part in the special commission in charge of the identification of unsanitary dwellings and of the causes of insalubrities, and the adoption of appropriate measures of cleanliness (art. 2).[8] It also gave the *Conseil Privé* the authority to review the appeals of the city residents to the renovation measures ordered by the municipal council (art. 3).[9] Housing legislation also focused on slum clearance and street alignments and enlargements, and targeted 'dangerous, insalubrious or incommode dwellings,' a clear reference to the thatch-roofed dwellings belonging to the urban poor, still standing in the city center between Saint-Pierre and Thionck Streets.[10] More data is needed to determine the number of dwellings that were affected by the law, the basis for the appeals, the number of people who complied with the decisions of the municipal council, and cases of non-compliance as well as cases prosecuted or fined.

## Contagionists and Quarantine Measures

Given the persistence of the yellow fever threat, contagionists pushed for more reliance on quarantines as a tool for combating the spread of yellow fever. In their construction of yellow fever epidemics, contagionists had identified Freetown (Sierra Leone) and Sainte-Marie-de-Bathurst (Gambia), both British colonial port-cities in West Africa, as the 'foci of infection.' Evidence needed to support their claims could be found in the works of their predecessors and in their own observations. Indeed, in his *Eléments de pathologie exotique* (1881),[11] Maurice Nielly had argued that yellow fever had already reached 'endemo-epidemic proportions' at the mouth of the Gambia River and in Sierra Leone. Another public health authority, physician Armand Corre, in his *Traité des fièvres des pays chauds* (1883), had claimed that Sierra Leone was 'the hotbed where most epidemics originated from.'

The outbreak of an epidemic of yellow fever in Dakar in November 1882, despite the imposition of the quarantine in July,[12] underlined the urgency of 'complementary precautionary measures.' Preliminary reports indicated that the epidemic had already provoked 20 deaths, including one physician, one engineer, and two railroad employees. The public health authorities in Saint-Louis imposed a sweeping 21-day quarantine on ships arriving from Dakar and its hinterland. No contact was allowed between Dakar and the outside world. Even ships arriving from Gorée and Rufisque were subject to seven-day observation quarantine at Bop-Diara. Sanitary guards were placed on Khor and Leybar Bridges to prevent contamination by land from people suspected of communication with contaminated regions. Copies of the law of March 3, 1822 related to sanitary police and of the ordinance of December 5, 1882 were printed and posted all over the city.[13] By mid-December, the epidemic subsided.

The danger of infection was heightened in early 1883 by confirmed reports that some ship captains arriving from the south with clean bills of health made frequent transits in 'intermediary places,' such as secondary trading posts, where there were no public health officials, thus making it difficult for the Chief Medical Officer in Saint-Louis to determine the real sanitary status of those trading posts and the validity of the clean bills of health in their possession. In order to deal with this situation of uncertainty and to protect the capital city from another epidemic, a new ordinance was passed on June 30, 1883 which gave the *Médecin arraisonneur* based at the quarantine station in Gandiole the authority to thoroughly inspect all ships from the Atlantic Coast destined to Saint-Louis, but those arriving from Europe whose inspection was limited to the evaluation of the responses of ship captains to the questionnaire related to the bill of health. Suspected ships were subjugated to a temporary quarantine pending the decision of the Saint-Louis Sanitary Commission. Contaminated ships were not allowed to navigate beyond the bar. During the rainy season, between 1 June and 15 December, ships arriving from Upper Senegal were required to obtain bills of health at Bakel and to be inspected either at Podor or at Richard Toll, the results of which were immediately transmitted by telegraphic cable to the Chief Medical Officer. Suspected ships were re-examined at the quarantine station of Bop N'Kior by a sanitary guard before being allowed free navigation and communication with Saint-Louis. During the inspection passengers were not allowed to leave the ship unless for compelling administrative reasons. Thus given that the Sanitary Commission had exceptionally broad powers, the possession of a clean bill of health and the absence of disease aboard were no longer sufficient conditions, even in an emergency situation, for a free navigation. Only the sanitary inspection of the ship by the *Médecin arraisonneur* had to determine its eligibility for a clean bill of health. In order words, the new legislation reactualized the instructions dated July 28, 1876, and the provisions of the ordinance of June 25, 1879, especially article 4 which imposed a five-day observation quarantine between 1 June and 15 December on all ships arriving from areas located between Pointe Sangomar and Gabon. Because of inertia or incompetence or even practical reasons, these sanitary measures were no longer strictly enforced.[14] The implementation of the new legislation would have a negative impact on the operations of commerce because of long delays that interrupted the free movement of people and goods and paralyzed commerce.

Although the anti-yellow fever measures adopted by the administration, including the construction of one lazaretto in Saint-Louis and another in Dakar, the recruitment of qualified sanitary personnel, and the quarantine measures, were financially costly, the authorities were convinced that such expenses were still far lower than the cost that high mortality rates, the paralysis of commercial activities, and the material losses resulting from a single epidemic could produce.[15] The lack of trust in the validity of bills of health led to the next logical step, that is, the imposition of an automatic annual quarantine.

Relying on the environmental paradigm, on June 19, 1884, the contagionists recommended the imposition of a strict quarantine of 23 days (including the navigation time) on ships arriving directly from Sierra Leone and a quarantine of observation of variable duration on ships arriving from 'countries in free communication with Sierra Leone,'[16] a clear reference to Gambia and other British colonial posts and their dependencies.

Having taken preventive measures against Sierra Leone, Gambia and their dependencies, the contagionists turned their attention to the other perceived dangerous region. Starting in August 1884 a five-day automatic observation quarantine was imposed every year during the rainy season between June 1 and December 15 on all ships arriving from the region between the Pointe de Sangomar (at the mouth of Casamance River) and Gabon no matter the status of the bill of health or the sanitary conditions of the area (art. 113 of the decree of August 29). The automatic observation quarantine was also imposed at Bakel on all ships arriving from Upper Senegal without clean bills of health; if not, ships were required to stop at the posts of Podor, Dagana, and Richard Toll, or if needed, at Bop N'Kior (art. 110). Exception was made for ships arriving directly from Gabon with clean bills of health issued by the French authorities if they had 'no compromising communication' with other ships or on land since departure (art. 114).[17] Another region of the world that was looked at with the same suspicion as a 'focus of infection' was the American coastal region between the tropics, including the islands (at the exception of the French Caribbean of Antilles and Guyane), where yellow fever was endemic (art. 31). The rationale behind this radical measure was that past epidemics that struck Senegal (1830, 1837, 1959, 1867, and 1878) originated from localities on the West African coast (mouth of Niger River, Sierra Leone, and Côte d'Ivoire).[18]

The response of the colonial state and city government to yellow fever threat in terms of revising many of the provisions of existing sanitary laws as well as adding new ones, was based on the assumption that the medical infrastructure was adequate, especially concerning the lazarettos, that merchants and ship captains were likely to cooperate with public health officials enforcing the quarantine regulations, and that commercial operations would be protected. But the implementation of the decree would prove those assumptions wrong.

## Protests against the Automatic Quarantine Measures

The attitude of the merchants toward the 1884 quarantine measures was a contributing factor to their failure. Indeed, merchants were rightly displeased with the interruption of all kinds of communication and exchange between colonial port-cities that the new measures created. The majority of French, British, and Portuguese merchants protested against the poor conditions of the implementation of the quarantine measures for economic and practical reasons. For example, on September 12, 1885 British merchants operating in Sierra Leone addressed a letter to V. Bareste, French Vice-Council in Freetown, in which they expressed their concern about

> the great inconvenience, loss of time and money, entailed upon us by the existing quarantine regulations in the Rivers to the North of this port. In view of the fact that there is no epidemic or disease of any kind prevalent (except the ordinary African fever, of which there are only very few cases, these cases confined to natives of Sierra Leone) we feel that your government would be justified in raising the quarantine.[19]

The 'ordinary African fever...confined to natives' was presented as a fever that was natural to the Africans and did not require any special medical attention, thus corroborating the idea that as long as a disease affected the natives, it did not receive a *registre civil* or official status. In any case, the delays were due to a certain number of factors. There were not enough buildings in the lazarettos and sanitary stations to accommodate different cohorts of passengers, who arrived at different dates, or to store and disinfect merchandise. There was shortage of medical personnel who could perform medical examination of the passengers, provide free health care for the sick, and disinfect merchandise; and there was also shortage of sanitary guards who could police lazarettos and sanitary stations. In addition, delays were also due to either the incompetence or the unwillingness of the bureaucrats and ship crews to accomplish the necessary tasks.

By 1889 the consensus among the authorities and the business community in Senegal and Paris was that the automatic quarantine was inefficient and a ruin for the State, the colony and commerce. As long delays in ports and cases of fraud were brought to the attention of the authorities, the urgent question before them was about how to conciliate the interests of commerce with those of public health. A look at the British colonies in West Africa revealed that British commercial interests prevailed over any other considerations. Indeed, there were reports that British physicians in Sierra Leone and Gambia were issuing fantasist and fraudulent clean bills of health where cases of yellow fever were given fantasist names that made them of little medical value.[20] Ship captains were seen disembarking their passengers at any place along the coast before entering the ports of Rufisque and Dakar to avoid medical inspections. At the same time, the toll of the quarantine on economic activities was becoming heavy, especially on the operation of the railroad Dakar-Saint-Louis inaugurated in 1885.

Accusations against the British complacency vis-à-vis yellow fever in West Africa were not without foundation. In India the British also showed the same kind of complacency in combating the threat posed by cholera. Indeed, being the champions of free trade, there too the British saw the quarantine system as a handicap to commerce and were hesitant to use it. Other reasons explained the British attitude, including the fear of rebellion, as cholera was associated with Hindu pilgrimages and festivals, the financial cost of the quarantine operations, police corruption, and the uncertainty of the medical knowledge concerning the etiology and mode of transmission of cholera. This was in contrast with the British response to the 1896 plague epidemic in Bombay when they took drastic measures to combat the disease threat.[21]

In any case, in search of solutions, the anti-contagionists among the French colonial authorities called for the suppression of the article 113 of the decree of 29 August 1884, the appointment of a Navy physician at Bathurst (Gambia) as *Médecin Arraisonneur* who would supervise the issuance of bills of health, and more involvement from the French Consuls in Sierra Leone and Gambia.[22] But despite the pressure from the merchants, the Council of Public Hygiene and Salubrity (CPHS) refused in May 1890 to allow the importation of kola nuts from Sierra Leone. Fear of the return of epidemics was not exaggerated because cases of cholera and yellow fever were reported in Spain.[23] The quarantine question was increasingly framed in terms of how to reconcile commercial and financial interests with the superior interest of public health. In a Sanitary Commission meeting held on 23 July 1891, the head of the Administration Service, de Korsaint-Gilly, suggested the suppression of the quarantine and its replacement by the disinfection of ships and trains coming from infected regions and the installation of lazarettos and sanitary cordons at the entrance of all uncontaminated centers. But the *Médecin en Chef*, Dr. Ayme, rejected the suggestion and argued that the colony of Senegal could not afford any fresh spending in new construction. At the end of the meeting, 8 out of 11 participants voted in favor of the suppression of trains in case of outbreak of an epidemic disease along the railroad.[24]

With time the Chamber of Commerce of Saint-Louis began to raise doubts about the scientific knowledge on which all these 'radical measures' and 'irrational practices' were based, when newspapers featured commercials about new disinfection machines 'capable of destroying the microbes of yellow fever in a few minutes at a temperature of more than 100 degrees.' Merchants mobilized to purchase without delay the disinfection machine 'no matter the sacrifice.'[25] Navy officers too contended that the quarantine was detrimental to the conduct of war in Sudan because it affected supplies to the troops.[26]

Despite the physicians' increasing role in shaping and maintaining the colonial order, the pressure from various chambers of commerce in Senegal was such that the authorities had to find another strategy. Even the outbreak of cholera in Saint-Louis in 1893, which revealed the deep social and economic disparities between sanitary citizens and unsanitary subjects as well as the inadequacies of public health policies (chapter 3), did not prevent Governor de Lamothe from recommending either the abrogation or, at least, the modification of the articles 113 and 114 of the decree of August 29, 1884, targeting ships arriving from the region between the Pointe de Sangomar and French Congo. But physicians, attributing the absence of yellow fever epidemics in Senegal since 1884, despite the outbreaks reported on other coastal trading posts, to the effectiveness of the annual automatic quarantine, strongly recommended the status quo. Finally, state officials, public health authorities, and the chambers of commerce agreed on a compromise solution: the modification of the articles in question by making provisions for delivery of clean bills of health, after boarding and examination of ships, unless suspected cases were observed aboard and information concerning the sanitary situation of the region seemed insufficient and insincere, in which case observation quarantine was imposed. The authorities had hoped that the modification of the sanitary legislation would allow French merchants to again compete effectively with British and Portuguese merchants operating on the West African coast, while protecting the colony against epidemics.[27]

A change in the sanitary legislation intervened in March 1897 when, in compliance with the resolutions of the 1893 Dresden International Sanitary Convention (15 April) (see chapter 3), promulgated in France in 1894 (decree of 22 May), another decree unified and consolidated different pieces of legislation applied until then in French colonies. The new decree reorganized the sanitary police, accountable to the Chief Medical Officer, which was responsible for the operations of disinfection at the quarantine stations or lazarettos and sanitary stations on coastal regions. Ship captains leaving the ports of residence were required to obtain bills of health indicating the sanitary condition of the point of departure. The decree also made compulsory the presence of a doctor aboard postal ships or any ships carrying 100 or more Europeans for more than 48 hours. Upon arrival in a port and before communicating with people in the port, any ship carrying a clean bill of health was visited by the sanitary authority and the captain interrogated under oath on health matters. If the ship was suspect, then it was submitted to a special sanitary inspection that included, depending on circumstances, a medical visit of passengers and the crew.

The decree also had provisions for the disinfection or isolation of suspect or contaminated ships. The compulsory disinfection concerned items, such as clothes, beds, objects, carpets, animals, and merchandise originating from contaminated regions. Other objects suspected susceptible of transmitting contagious diseases, such as animal skins, were simply destroyed. A quarantine of observation was imposed upon any ship that had left a region ravaged by cholera less than seven days earlier, and a country ravaged by yellow fever or plague at least nine days earlier. The quarantine lasted seven days for cholera and nine days for yellow fever and plague at the lazaretto for passengers without sanitary passports, the unloading and disinfection of merchandise, and the disinfection and quarantine of ships.

Sanitary councils would be created in the ports open to trade in order to represent local interests in the decision-making process related to health issues. The members of these councils would represent various administrative, military, scientific, and commercial interests. The decree also had provisions for the creation of a Sanitation Committee in the capital-city of each colony as well as a Sanitation Commission in other places in order to promote public health and general hygiene. It required passengers to obtain sanitary passports.[28]

Those who contravened these measures would be punished according to sanctions contained in the law of 3 March 1822 related to the sanitary police: death penalty for those who would communicate with contaminated regions; life prison and between 200 and 2,000 francs in fees for communicating with countries under the regime qualified as 'suspect'; from one to ten years in prison and from 100 to 10,000 francs in fees for a prohibited communication with places, people or things being quarantined.[29]

The importance of these drastic sanitary measures also lies in the fact that the bureaucrats in charge of implementing the new policy could also face severe sanctions in case of failure on their part. For example, death penalty was requested for a sanitary agent, a state ship agent or a private company

> who, officially, in a telegram, a certificate, a report, a declaration or any evidence, would alter or distort the facts, so as to compromise the public health...if a pestilential invasion followed.[30]

The same sanction also applied to any individual working in a sanitary cordon or any person affected to the quarantine that would abandon his post or violate the instructions. If the disease did not spread, such an agent would face forced labor and a fine between 10 and 20,000 francs. Other sanctions could apply to various cases concerning the refusal to implement sanitary measures or the neglect of duties. Thus, the obligation for a ship, no matter its origin, to show its bill of health upon arrival in a port of a French colony or a region under French protectorate was motivated by the need to protect public health.

The success of the new decree depended on renovations in sanitary services in terms of sufficient personnel and equipment. But in practice, the implementation of the decree revealed problematic. The medical personnel were insufficient, and the medical infrastructure was inappropriate and, in many coastal trading posts in Côte d'Ivoire, French Guinea, Dahomey, and French Congo, very embryonic. The administration of the newly created government of French West Africa (1895), based in Saint-Louis, lacked adequate financial resources needed to build isolation camps or lazarettos and sanitary stations, and to equip them with infirmaries, separate barracks for accommodating the sick and the suspects, separate buildings for merchandises and other objects to be disinfected or already disinfected, safe drinking water, a sewage system, and machines for disinfection. Also the decree rested upon questionable assumptions about the bureaucratic process and the interests at stake.

The majority of merchants, however, voiced their opposition to the new quarantine measures they described as 'inefficient and absurd,' in part because of the deficient organization or the underdevelopment of health services. The Chamber of Commerce continued to recommend the acquisition of disinfection machines as a solution to 'the costly and unpopular quarantine.'[31]

## A Crisis of Confidence

Following reports of the outbreaks of yellow fever epidemics on May 2, 1899 in Côte d'Ivoire and Gambia, the Sanitary Commission imposed quarantine on suspect or contaminated ships arriving from Côte d'Ivoire. The decision provoked more discussion concerning the merit of the quarantine system per se.[32] What increased the anxiety was the fact that the news spread only after the *'Stamboul,'* which had arrived from Côte d'Ivoire, had departed with a clean bill of health after spending a few hours of transit in the port of Dakar. Thus, fear of infection had greatly contributed to the malaise. But Minister of Colonies Albert Decrais, increasingly under the pressure from merchants who pushed for more rational sanitary measures informed by the changing medical knowledge about the etiology of infectious diseases, on 29 May, cabled Governor General Chaudié requesting that the Sanitary Commission examine the possibility of allowing the African Steamship Company fleet to enter the ports of Côte d'Ivoire not contaminated by yellow fever, given that yellow fever cases were reported only in Grand-Bassam. The Sanitary Commission met on June 2 to analyze the situation. In his explanation of the causes of the recent disease threat, Chief Medical Officer Lafage attributed the re-emergence of yellow fever to the compromised solutions adopted since 1893 under pressure from the business community. He argued that the surveillance system of automatic annual quarantines inaugurated in 1884 had succeeded in protecting Senegal against yellow fever attacks and that its suppression increased the colony's vulnerability, and that 'the return of epidemics would be a calamity and the ruin for the colony.'[33] Dr. Lafage also found the request from Minister of Colonies Decrais unacceptable based on the fact that the absence of a regular medical service in these localities of Côte d'Ivoire made it difficult to determine the presence or absence of yellow fever. During the final vote the Sanitary Commission rejected the Decrais' proposal based on the argument made by Chief Medical Officer Lafage.

Fearing the 'incalculable economic and social consequences of an eventual epidemic,' Governor Chaudié warned Minister of Colonies Decrais that 'in Senegal, yellow fever epidemics were always so serious that it was necessary not to neglect one single precaution to prevent the disaster from entering the colony,' thus re-affirming the credo of the toxicity of the climate of Senegal, its 'tropicality.'[34] But noting that the authorities in Dahomey had already applied precautionary measures and provided clean bills of health to ships from ports of Côte d'Ivoire other than Grand-Bassam, Decrais urged the same day by cable the authorities in Senegal to follow the Dahomean example. This time, the Sanitary Commission not only stood its ground but also overwhelmingly endorsed Dr. Lafage's proposition to revive the article 113 of the decree of August 29, 1884 that was modified in 1893. In other words, the Sanitary Commission suggested a re-imposition of the five-day automatic annual observation quarantine to ships arriving in ports of Gorée, Dakar, Rufisque, and Saint-Louis from regions located between Pointe Sangomar and Gabon between 1 June and 15 December, even if they held clean bills of health, thus re-affirming the status quo ante. Thus, this exchange revealed the ongoing opposition between contagionists (here the Sanitary Commission and Governor General Chaudié) and anti-contagionists (merchants and Minister of Colonies Decrais). But this was not the end of the debate. Four days later, the question about how best to protect Senegal against the much-feared yellow fever was again discussed.

During another debate of the Sanitary Commission of June 6, 1899 in Saint-Louis, presided over by Th. Bergès, secretary general of the government, the discussion again centered on finding the best solution to protect Senegal against the outbreak of yellow fever in Grand-Bassam without penalizing the other ports in Côte d'Ivoire where there were no yellow fever cases reported. The participants were faced with two contradictory demands. One came from Minister of Colonies Decrais who reiterated a previous proposal that was already rejected by the Sanitary Commission four days earlier, urging that a clean bill of health be provided to ships arriving from city-ports of Côte d'Ivoire other than Grand-Bassam after taking the necessary precautions, as it was the case in Dahomey. The other proposal was presented by Chief Medical Officer Lafage who opposed such 'complacency.' In a report to Bergès, he had argued in favor of the re-instatement, until further notice, of article 113 of the decree of August 29, 1884. Dr. Lafage's rationale was that the lack of sanitary surveillance outside Grand-Bassam as well as the absence of sanitary agents on the coast of Liberia did not allow a close monitoring of the spread and prevalence of the 'disaster' along the West African coast. He warned that we will suddenly learn that it just broke out in Sierra Leone, in Guinea, in Casamance, in Bathurst. We will then be close to being caught and it will probably be very late to take measures the efficiency of which we can rely on.[35]

Dr. Lafage's fear of another epidemic was justified not only by the presence in Dakar/Gorée area of 2,000 European soldiers who were living in defective conditions at the approach of *hivernage* but also by the evidence of unreliability of the statements made by ships' captains arriving from the south. Dr. Lafage's theory was that yellow fever was now endemic along the coast of West Africa, which caused the potential for epidemic flare-ups. In his view, 'the epidemic will continue its ascendant march toward the north and it is when it will appear extinguished, in places where it is now ravaging, that it will break out in other places.' He presented the evidence from the 1881 yellow fever epidemic, which 'started in Gorée in a store where clothes were locked up for three years' to make his point.

Bergès approached the question in legal terms. From his perspective, the public health authorities were not allowed to impose quarantine on ships arriving from countries where no 'foyers of infection' were reported and that were 'distant by several days of navigation from places where the epidemic is declared.' Thus, the ports of Côte d'Ivoire other than Grand-Bassam fell under this category.

The majority of participants expressed concern over the provision of article 113 of the decree of 29 August, 1884 related to a careful disinfection of passengers, luggage, and merchandise; they found that such a measure could negatively affect economic activities and result in the payment of damages to merchants. In the face of such a resistance, Dr. Lafage proposed to limit the operation of disinfection to the objects and cloths used by the passengers; and the measure was adopted.[36] This outcome underlined the increasing skepticism about the contagionist theory that informed the disinfection credo. In addition, by compromising on the key principle of disinfection of merchandise, Dr. Lafage sent the wrong message from the point of view of contagionists that medical knowledge was negotiable, contingent. Another conclusion that can be drawn from this outcome is that the power relationship within the Sanitary Commission was increasingly shifting in favor of commercial interests. It was a short-lived victory for the anti-contagionists who now dominated the Sanitary Commission, however, since their recommendation had yet to face the scrutiny of the *Conseil Privé*.

In the meantime, as his first cable remained unanswered, the next day (June 7), Minister of Colonies Decrais in Paris cabled Governor General Chaudié a reminder at the time when the latter had just received a cablegram from Governor Cousturier of Guinée informing him that the last reported yellow fever case dated 12 days earlier. In order to provide a comprehensive response to Paris, Governor

General Chaudié hastily held on 8 June a special meeting of the *Conseil Privé*[37] that was extended to the four mayors of Senegal and to the representatives of the army (Colonel Pujol), the Saint-Louis Chamber of Commerce (Fr. Rabaud), and *Conseil Général* (G. d'Erneville) aimed at evaluating the sanitary situation of the colony. The participants were asked to examine two crucial questions: 1) the continuation of the quarantine measures dictated by the decree of March 31, 1897 that were applied to ships arriving from Côte d'Ivoire, 2) and the examination of the Chief Medical Officer Lafage's proposed re-instatement of article 113 of the decree of August 29, 1884. In his opening statement, Governor General Chaudié invited 'any person well knowledgeable about the needs and desiderata of the population to interpret (for the audience) the *public sentiment* concerning the quarantine measures adopted so far or to be adopted in case of an epidemic' (emphasis added). In response, Raymond Martin, representative of Buhan and Tesseire Company in Saint-Louis and Dakar, affirmed that the population's response to the suppression of (traditional) preventive measures was negative and that the current measures were viewed as 'insufficient' in the light of the physicians' inability 'to quickly declare the existence of the affliction.' Using the evidence from the medical events surrounding the outbreak of the 1878 yellow fever epidemic, he explained that the fact that physician Massola and his colleagues could not agree on the result of the autopsy performed on attorney Batut left them powerless in face of a growing danger that required immediate preventive actions, since, in his views, it took a month and half for yellow fever to incubate. Martin's account reiterated the epidemiological theory developed by Bérenger-Féraud in his *Traité théorique et clinique de la fièvre jaune* (1891) (see chapter 2).

Dr. Lafage corroborated Martin's account of the epidemiology of yellow fever, using the evidence from the 1881 great yellow fever epidemic. According to his theory, yellow fever broke out after Dr. Jullien ordered the cleaning of the office next to his office that contained a suitcase belonging to a *tirailleur* who had succumbed to yellow fever in 1878. In the process of opening the suitcase, Dr. Lafage's story went, Dr. Jullien was 'struck by the terrible affliction to which he soon succumbed.' Chief Medical Officer Carpentier also died from the disease, which re-emerged in Gorée. Following Dr. Lafage, Martin added that the 1881 epidemic broke out 'with an extraordinary intensity' after Governor de Lanneau initiated the cleaning of some storage rooms located on the first floor of the *Hôtel du Gouvernement*. The governor, all officers, and almost all civil servants working in the building succumbed to the disease shortly afterwards. Martin concluded that, given that the beginnings of the disease were difficult to diagnose and that contagion through objects was 'tenacious and durable,' it was indispensable to take 'severe measures to prevent the disaster from entering Senegal.' Thus, both Martin and Dr. Lafage repeated one of the basic tenets of contagionism concerning the mode of transmission of yellow fever that stressed contact with infected objects confined in closed spaces that were untouched since the last epidemic, or even houses where yellow fever victims had lived.

In his response to Martin's and Dr. Lafage's comments, Governor General Chaudié, who appeared scientifically minded and sympathetic to the interests of commerce, argued that in recent years 'science has made progress and that in fact, the decree of 31 March 1897 has provisions that the highest medical authorities find efficient enough to prevent and stop contagion.' But he also recognized that the key to the success of preventive measures depended on the ability of the Sanitary Service 'to assure their total implementation either at the point of departure in the contaminated country, or en route, or and mostly at the point of arrival.' Bergès contended that the embryonic state of the Sanitary Service in Côte d'Ivoire rendered the clean bills of health delivered by the medical authorities there untrustworthy. But Governor General Chaudié insisted that, given that *hivernage* facilitated the spread of yellow fever, 'now is the time for us to be stricter.' Following this exchange, the participants unanimously voted to uphold the quarantine measures imposed on ships arriving from Côte d'Ivoire despite the absence of new yellow fever cases in the past 12 days. They also approved the content of the cable to be sent to the Minister of Colonies in Paris, as follows:

> The Sanitary Commission, the *Conseil Privé* and all the Mayors of the Colony, despite the news brought to our attention by your cablegram dated 1 June, are of the opinion that because of the season specially favorable to the spread of yellow fever, it is absolutely impossible to provide clean bills of health to any ship that has communicated with Côte d'Ivoire. In any case, the *public sentiment* unanimously acquiesces with this opinion (emphasis added).[38]

If the participants easily agreed on the necessity of quarantine measures against Côte d'Ivoire, they sharply disagreed concerning which of the two decrees under discussion (1884 and 1897) would be more effective in preventing yellow fever from reaching Senegal. The opposition to the re-enactment of article 113 of the decree of August 29, 1884 was led by Fr. Marsat, a pro-commerce Mayor of Dakar, who objected to any move to re-enact the 'draconian decree,' and recalled the long struggle led by the chambers of commerce and municipalities to convince Chief Medical Officer Ayme to abolish the quarantine measures that constituted a tremendous handicap for the freedom of commerce. The opposition camp did not see the need to impose the five-day observation quarantine on ships arriving from uncontaminated regions and to deny clean bills of health to ships arriving from Dahomey and Côte d'Ivoire. Instead, they were in favor of the strict implementation of the decree of March 31, 1897, which was approved by the Sanitary Commission and the Chief Medical Officers, took into account the incubation period for yellow fever, and had provisions for the disinfection of passengers and merchandise (art. 55).

The most outspoken member of the camp that favored the re-enactment of article 113 of the decree of August 29, 1884, was Raymond Martin who, in his capacity as a representative of a major company that had big business interests in the Dakar region, contended that sanitary measures imposed on ships arriving from the south were not harmful to commerce in the Dakar region. He rejected the quarantine measures outlined by the decree of March 31, 1897 as 'insufficient.' He called on the participants to be 'more rigorous toward the bizarre manifestations of yellow fever which, as we have seen, often take a very long time before declaring themselves,' and he insisted that in special circumstances like this one '*commerce must sacrifice its interests* (emphasis added). Coming from a representative of a major business company, such a statement was truly remarkable.

In the light of the increasing tension, Chief Medical Officer Lafage suggested that the re-enactment of article 113 be limited to the current *hivernage*. At the end of the debate, the majority of participants voted by 9 voices in favor and 3 against the re-enactment of the article under discussion.

Governor General Chaudié emphasized the temporary character of the measure and read the second part of the cable prepared for Minister of Colonies Decrais, as follows: 'These authorities asked me, in contrary, to revive, temporarily and for the current hivernage, the provisions of the decree of August 29, 1884 which imposed five-day observation quarantine to any ships arriving from the South. I request your instructions.' The cablegram was adopted unanimously.

By emphasizing the temporary character of the new quarantine measures and by dissociating himself from the text just adopted ('these authorities asked me'), Governor General Chaudié expressed his displeasure with the attitude of the 'radical' contagionist members of the *Conseil Privé*. This fact did not go unnoticed by R. Martin who, after the adoption of the cablegram, requested that a sentence be added to the text indicating that 'the Head of the Colony associates himself with the request formulated by the majority of the Assembly.' In response, the governor general took great pain to explain that he had the duty not to express his opinion when it was question to re-visit the more severe provisions of an act whose abrogation was done only after mature examination and on request from the most competent and most qualified authorities for the preservation of the Public Health in the colonies as well as 'at home.' Nobody requesting the floor, the one-hour-and-half session was terminated.

Governor General Chaudié was convinced that the decision to re-impose the provisions of the decree of August 29, 1884 was 'inspired by the fear, even by the terror, that people in Senegal have of the possibility of the re-emergence here of a yellow fever epidemic.' In a letter to the minister of colonies, dated June 15, 1899, Chaudié explained that

> no measure seems severe enough or draconian enough to them (contagionists) to the point that Mr. Raymond Martin, *conseiller privé* for fifteen years and certainly one of the most enlightened minds in Senegal, did not hesitate, along with the large majority of participants, 9 versus 3, to insist on the return to the implementation of the provisions of the decree of 29 August 1884.[39]

Once informed, the minister of colonies, who clearly defended the interests of commerce, reversed the decision made by the *Conseil Privé*, however, and instructed the governor general to impose the quarantine only on ships arriving from Côte d'Ivoire, while urging the *médecin sanitaire* of Dakar to be more vigilant with the medical examination of the arriving passengers and the disinfection of merchandise.[40]

Defenders of commercial interests won a major victory. They demonstrated that they had friends in high places in Paris, and the growing influence of the 'Bacteriological Revolution' strengthened their arguments. So they continued to put pressure on the administration officials both in Saint-Louis and in Paris in order to push them to rethink their public health policy aimed at protecting the colony from yellow fever that broke out in Grand-Bassam. Contagionists still had sympathizers within the General Inspection of Health Service in the ministry of colonies in Paris, however, who believed that as long as French colonies south of Senegal (Guinée, Côte-d'Ivoire, Togo, Dahomey, Cameroun, Gabon and Congo-Brazaville) did not implement sanitary measures prescribed by the 1897 decree, the medical authorities in Senegal were entitled to declare all ships arriving from there suspect. From the point of view of contagionists, the central administration in Paris would be equally responsible for any future 'disasters' if they did not put more pressure on the local colonial authorities who were slow to implement the provisions of the 1897 decree.[41] Opposition to anti-yellow fever measures increased when yellow fever returned to Senegal the following year and the quarantines were again imposed and the daily life in Senegal was brought to a standstill.

## Protests Against the 1900 Anti-Yellow Fever Measures

The raw emotions to the news of the outbreak of yellow fever, as seen in chapter 2, were followed by drastic measures against the epidemic disease. The office of the governor general was flooded with complaints from all over Senegal and French

West Africa over the negative impact of the quarantine and sanitary cordons on food supplies and prices of commodities. Thirty-eight 'Europeans and inhabitants of Tivouaouane' protested against the suspension of the delivery of ice made in Saint-Louis described as 'a precious resource for the society' and 'the most precious auxiliary for the doctor during the rainy and hot season.'[42] After unsuccessful 'repeated petitions' that attacked the governor general's measure 'to continue to authorize (the) shipment of merchandise from contaminated places to still uncontaminated train stations,' the 'Population of Tivaouane' solicited the intervention of the minister of colonies to 'stop the development of the epidemic.'[43] Inspector General Kermorgant approved their request and ordered the interruption of all communications between contaminated commercial firms and places still devoid of disease, interruption that also 'indistinctly applied to people, animals, and all kinds of merchandise.'[44] Governor General Chaudié complied but his decision was met with more protests from Dakar.[45] Even merchants who had left Saint-Louis in early May began to worry about the situation of their businesses left behind.[46]

Protests continued throughout the summer months. Traders from the hinterland of Dakar, Rufisque and Saint-Louis, who were directly affected by the quarantine measures, voiced their opposition to such measures they found 'radical.' In a strongly worded letter to the governor general, dated September 7, 1900, M. Meyer, representative of the *Société le Syndicat du Soudan Français*, who was blocked in Saint-Louis with 100 tons of merchandise from France, protested against the rigorous quarantine which resulted in a great loss for the traders. He wrote that he could not find one example in Europe or elsewhere in other colonies, where commerce was completely stopped because of the epidemic. He emphasized that the quarantine intervened between August and September, at exactly the time when the Senegal River was navigable to Kayes.[47]

The quarantine measures prohibiting communication with trading posts along the Senegal River also provoked tension between Governor General Ballay and the *Commandant Supérieur des Troupes* Combes, as the quarantine measures prevented Paris from sending fresh troops to French West Africa to fight African resistance to colonial intrusion until the end of the epidemic, thus forcing the Commandant to face the 'requirements of the exceptional situation with reduced resources in personnel at his disposal.'[48] With the remaining 69 soldiers and 3 members of the Joint Chiefs of Staff (*Etat Major des Troupes*) out of 16, besides the 3 indigenous sentries, General Combes found himself 'in the impossibility to fulfill his duties.'[49] He remained opposed to the sanitary measures even if they were adopted in order to protect the lives of the European military personnel. He questioned the governor general's decision not to let the ship '*Hirondelle*' make the trip to Kayes for fear of spreading the disease across Upper Senegal because of prior reports of yellow fever cases that were declared aboard.[50] Thus, Governor

General Ballay did not see General Combes as an ally engaged in the same struggle to protect European lives.[51]

The local elite also opposed the quarantine. During the meeting of the Colonial Commission of July 28, 1900, Justin Devès shifted emphasis from economic, administrative, and psychological reasons to medical reasons for the suppression of the quarantine. He contended that physicians were helpless against a disease 'that escapes modern science and empirical processes,' and that yellow fever affected only a minority of city residents (Whites) who were 'essentially vulnerable,' while the majority was either unscathed or refractory to the disease, but it interrupted the economic and social life. Instead of sanitary cordons and quarantine, so the argument went, the best solution would be to provide White colonists with medical leave and/or home care, and to disinfect suspected houses, which suggested that the *métis* and Blacks also approached yellow fever as the 'White man's disease' that did not affect them, thus reflecting the official thinking concerning racial susceptibility. The Municipal Council also shared the same views and recommended that the sanitary measures affecting trade be implemented with 'wisdom and prudence,' by allowing foodstuffs for the *indigènes* (rice, millet, couscous, sugar, cookies, oils, etc.) to be delivered by bi-weekly trains as it was the case for European supplies, such as bread, ice, potato, onions, wine, and canned food. By then 2,000 Europeans had already left for Europe.[52]

Five French merchants from Kelle, a train station between Saint-Louis and Dakar, in a letter to the minister of colonies also energetically protested against the carelessness and abuses of power shown by Governor General Chaudié, who made exceptions to the rules in favor of his cronies, allowed Mr. Heldt, owner of a buffet place, to violate the sanitary cordon in order to let his wife prepare a warm lunch for Madame Chaudié, who was fed up with the situation in Senegal. They mentioned other instances of misbehavior, including Chaudié's injunctions to the physician who denied Heldt access to the supposedly non-contaminated section of the train station; Mrs. Chaudié's attempt to enter by force into the 'Stamboul' after debarking from the train before it even reached the train station, for she did not communicate with the city. 'These facts,' the five French merchants wrote, 'well known to everybody here, were so grotesque that they will remain in the annals of the city.' The signatories also attacked the 'unqualified conduct' of governor general who, after establishing lazarettos at Kelle, Tiaroye, and Sebikotane, thus 'completely blocking Dakar' on 10 June, neglected to do the same in Saint-Louis and, in doing so, exposed all localities along the railroad where troops from Dakar and Rufisque were dispatched. Only after countless protestations and petitions did he accept to suspend the trains on Dakar-Saint-Louis railroad. Thus, the general perception was that Chaudié, by allowing only 'sanitary trains' to distribute foodstuffs only to troops, was not protecting 'all the French people living in the colony of Senegal,' including civilians, *négociants*, and railway workers, 'who were without bread since the 22nd' when all trains were canceled.[53] Chaudié was eventually dismissed as governor general of FWA for neglecting to implement official directives from the minister of colonies and for sending either incomplete or misleading reports to Paris.[54]

Sensitive to the pressure from various constituencies and in conformity with contagionist ideas, the authorities adopted measures that allowed the sick to isolate themselves in their houses instead of the hospital and provided specific guidelines for dealing with yellow fever: the isolation of the patient inside the house under the surveillance of a sanitary agent; the disinfection, with a biochloric or phenic solution, of parents and friends who were in contact with the sick, including their clothes; the disinfection of the furniture; the use of disinfectants in the vases receiving vomits and fecal matter; the compulsory report of deaths; a speedy burial of the dead inside the coffins; and the burning of the bedclothes, pillows and mattresses at the extremes north and south of the city under the surveillance of the police.[55]

The Dakar-Saint-Louis railroad workers also sent a petition to the minister of colonies about 'facts of exceptional gravity' involving exposure to infection and the lack of food supplies.[56] But, despite the pressure from the merchants, the quarantine remained in place until 2 December.[57] Calls were then made to merchants to send back their personnel to Senegal.[58]

## Bacteriology and the Control of Yellow Fever

The above discussions surrounding the adoption of anti-yellow fever sanitary measures show that by the 1880s and 1890s there was an increasing skepticism against the contagionist orthodoxy that had been the foundation of control measures against epidemic diseases. Constant references made to 'irrational practices' and microbes during various health boards' meetings discussing the usefulness of quarantines referred to the new medical advances made in the areas of infectious and parasitical diseases from the 1870s on by Louis Pasteur in France, his German rival, Robert Koch, and their followers, especially the knowledge that specific human and animal pathogens were caused by specific microorganisms or germs, and that there was a connection between a parasite or a virus, a vector (insect, mosquito, and rodent), and a human being or an animal; and the strong belief in the treatment or the prevention of infectious diseases. New institutions were created, including the Institut Pasteur of Paris in 1888, where the study of the virulence of microbes was under way, and schools dedicated to tropical medicine in Liverpool and London in 1899. There were also plans to open the Institut Pasteur in the colonies.[59] In this framework the Bacteriological Laboratory of Saint-Louis opened in 1896 and began its work before being transferred to Dakar and replaced by the Institut Pasteur of Dakar in 1924.

The new knowledge did not go unchallenged, however. Ample illustration is provided by the statements made by different protagonists during the debates over the quarantine measures. As Marc Renneville has shown, some physicians in France continued to contest the contribution of microbiology to medical knowledge. For example, Sigismond Jaccoud, Chair of internal pathology at the faculty of medicine until his death in 1883, found nothing new in the knowledge generated by microbiology. In his views, it was just a 'change of words,' microbe having replaced '*contage*' or miasma, which meant that the contagionist conception of disease control through strict quarantines and disinfection had to remain unchanged.[60] But Louis Pasteur also had many supporters within the medical establishment. One of them, Henri-Marie Bouley, general inspector of veterinary schools and member of the French Academy of Medicine (1868–1885), argued in favor of the incorporation of microbiology in the training programs of physicians in order to eliminate future physicians' resistance to the germ theory, the teaching of hygiene, and the political organization of public health. Another supporter of Pasteur was Jules Rocard, general inspector of health services in the Navy, who spent considerable energy to try to convince the medical community to embrace Pasteur's discoveries and to get politicians involved in the work of sanitary reform. In 1887 he argued against the contagionists' (or sanitarians') 'tyranny,' 'harassment,' and 'intransigence.' He contended that public health was negotiable and that competing financial, commercial and industrial interests should be taken into account when making health decisions.[61] Thus, echoes of the discussions at the meetings of the *Association Française pour l'Avancement des Sciences* (AFAS) (French Association for the Advancement of Sciences), where Pasteur, Koch and other scientists presented their findings and responded to their critics, can be found in the deliberations taking place in Saint-Louis.

Concerning the spread of yellow fever more specifically, new developments were taking place in Havana starting in June 1900 at the same time yellow fever struck Saint-Louis and Dakar and created panic. Indeed, the United States Army Commission, led by American bacteriologist Walter Reed and including American physician James Carroll, Cuban physician d'Aristide Agramonte, and American bacteriologist Jesse W. Lazear, began its investigation related to the etiology, spreading, and control of yellow fever. Inspired by Ross' findings on malaria, they started by testing Finlay's theory, using the eggs of *Aedes aegypti* provided by Finlay himself. Walter Reed presented their early finding at the American Association of Public Health meeting. The Commission found that the specific infectious agent of yellow fever could be transmitted to a non-immune individual through blood injection and produce an attack of yellow fever. It also confirmed 'what had so often been noted in the literature,' that yellow fever was not a contagious disease, and that it could not be transmitted by a convalescent patient directly nor through the objects that had been in contact with the patient, nor 'by means of fomites.' The

Commission also resolved a problem that had for a long time escaped scientists and medical authorities alike, that is, the 'bizarre manifestations of yellow fever,' which propagated in some places and not in others with the same conditions. Drawing from recent findings about the role of 'a special genre of mosquito' in the propagation of malarial fever made by Ross, Grassi, Bastianelli, Bignami and others, and from their own observations in August 1900 in Havana, the Commission established that yellow fever was transmitted by means of the bite of infected mosquitoes, called *Stegomya fasciata*. It rejected the received wisdom about the mode of transmission of yellow fever through the contact with contaminated clothing and bedding. Concerning the method of prevention of yellow fever, the Commission established that the best method was the destruction of mosquitoes with the fumes of pyrethrum and the protection of the sick against mosquito bites through isolation in a room equipped with wire screens. Drawing from these experiences, the sanitary authorities in Cuba undertook in 1901 to rid Havana and Santiago de la Vegas of yellow fever.[62]

It is true that the Walter Reed Commission's findings as summarized here were not readily available to public health authorities in Senegal when they were combating yellow fever during the summer months and in the fall of 1900. Its observations concerning the propagation of yellow fever by means of mosquito bites were made in August, while its Special Experimental Station was established toward the end of November and it continued its experiments until March 1901. Nevertheless, the new medical knowledge that was already available in the literature did not support the costly practice of disinfection of merchandise and other objects that had been in contact with the sick; the incineration of suspected items, as well as the quarantine system in cases of outbreaks of yellow fever; given that yellow fever was not a contagious disease. The new discoveries also made the policy of repatriation of non-essential personnel obsolete. But in Saint-Louis a combination of factors help explain the return to the status quo ante, including commercial interests, financial constraints, competing conceptions of public health visible in the physicians' skepticism about the Pasteurian revolution and the persistence of both the contagionist notions and the environmental paradigm—as seen in statements made by Drs. Ayme and Lafage above—as well as the epidemic of fear accompanying the outbreaks of yellow fever in Senegal.

## Toward a New Paradigm

By November 1900, opposition voices to the quarantine regime became louder and more forceful, even pressuring the public health authorities to learn from the experiences undertaken elsewhere. The administrator-director of the *Compagnie*

*Française de l'Afrique Occidentale* (CFAO), a company established in 1887 and headquartered in Marseille (France) that did business in West Africa, is a good case in point. In a letter to the minister of colonies, dated November 3, 1900, he laid down a rationale for the adoption of a different strategy. It is important to present his argument in full. He started by recognizing that the bad reputation of West African colonies resulting from the recurrence of epidemic diseases helped explain the hesitation the European capitalists and colonists had to make long-term investments in economic activities in a region 'so insalubrious.' Part of the evidence to support his claim came from Côte d'Ivoire where a year earlier 'bubonic plague decimated the *indigènes*, while yellow fever killed half of the European population,' thus reiterating the received wisdom about the racial susceptibility to specific colonial pathologies. Another piece of evidence came from Senegal where 'not long ago cholera ravaged; this year, it is yellow fever again, whose lamentable and persistent ravages you know of.'[63] He provided a long list of the most important 'immediate disastrous consequences' of the ravages of yellow fever epidemics, including the disorganization of public services, the repatriation or dispatch of troops, Navy shipyards coming to standstill, the cancelation of trains and ship schedules, the closing of businesses, the isolation of cities, the general exodus of White residents ('the only known efficient remedy to the disaster'), and the interruption of daily activities in the colony. In his view, the prosperity and development of Senegal were at stake since such high mortality and the sheer financial loss due to yellow fever, as well as the lingering psychology of fear, were likely to reduce the 'spirit of enterprise' among the colonists and to discourage potential colonists. In the eyes of outsiders, the bad reputation of Senegal as a diseased land was likely to impact on the whole West African coastal region. He feared that, if nothing were done, major public works, such as the construction of railroads and ports that were either at the feasibility study level or under way, could not be continued without the assistance of a great number of European personnel, whose recruitment was becoming harder and harder because of the fear of epidemic diseases.

Having underlined the lethal character of yellow fever and its economic consequences for the future of the colonial rule in Senegal, the administrator-director of the *Compagnie Française de l'Afrique Occidentale* recommended the adoption of 'measures based on the experience acquired in other regions.' More specifically, he pointed out that the reduction of the incidence and prevalence of epidemic diseases 'in some unhealthy regions of Europe and America' occurred thanks to the

> rational sanitation, the drainage of the soil, the provision of running water in sufficient quantity for domestic consumption and sprinkling in the streets, the construction of drainage and sewage systems and to the adoption of many other hygienic measures advocated by experimental science.[64]

He expressed his deep conviction that the 'same improvements, the same principles of sound hygiene, when implemented in Africa, will give, without doubt, the same good results.' At the same time, he assured the minister of colonies that the costs of the recommended innovations, designed to meet the needs of the European colonists, would not be too prohibitive for the budget of each colony in question. He also urged the minister of colonies to initiate soon 'a full inquiry in order to determine the causes of ill-health and the best means and practical way to stop its spread, and, possibly, to prevent its return.' He expressed his hope that the public health authorities and the Institut Pasteur would pursue their study of tropical diseases. Another recommendation focused on yellow fever in particular and centered on dispatching a mission, made of physicians and nurses, to regions where yellow fever was endemic, such as Havana, Mexico, and Brazil, and in hospitals where it was treated all year long in order to learn from 'the experience and local traditions...and curative methods, whose efficacy seem established.' The mission would decide about whether to recommend, 'even to impose, the use of crematorium furnaces, during the epidemic, and even, all the time, for the suspected cases...one (crematorium furnace) in each of the (French) colonial centers.' Anticipating that the minister of colonies could reject these recommendations because of financial constraints, the administrator-director suggested the use of local budgets in the colonies to support the expenses related to the mission. The final recommendation related to securing the vote for funds for public works (sanitation) from the elected bodies in Senegal and the Administrative Councils in French Guinée, Côte d'Ivoire and Dahomey. He concluded with an emphasis on 'the so considerably vital, humanitarian, and urgent interest for all these colonies, all these growing cities, which would not hesitate to make necessary sacrifices' in achieving these goals.[65]

The sense of urgency exhibited by the administrator-director of the *Compagnie Française de l'Afrique Occidentale* was justified by the fact that all decisions to send European troops back to Senegal were postponed indefinitely and yellow fever continued to produce victims even beyond *hivernage* in early January,[66] despite Governor General Ballay's insistence that these were isolated cases.[67] This situation of uncertainty reinforced the negative images of Senegal as the most unsanitary place in West Africa. Yellow fever continued to be constructed as the 'White man's disease,' which subsided with the repatriation of the Europeans and re-emerged with their 'premature debarkation' in Senegal, as they provided the disease with 'new food' (*nouvel aliment*).[68] In accordance with this perception, orders were given to delay the embarkation at Bresy (Bretagne) of two companies of Navy infantry and at Bordeaux of the personnel who were ready to leave for Senegal.[69] But the truth is that, despite the general panic, not all the Europeans left Senegal during the evacuation. Out of 98 agents of the *Compagnie du Chemin de Fer de Dakar à Saint-Louis*, 5 succumbed to yellow fever and 21 agents stayed in Senegal on their own demand. Pinaud, head of the Dakar train station who had joined his post in 1885, fell sick, recovered and resumed his functions.[70] In the absence of statistics, it is not possible to generalize from these minor cases. The majority of the colonists had left for France in panic at the beginning of the epidemic.

The authorities in Paris continued to be concerned with the spread of yellow fever and increasing mortality in Senegal, as well as the situation of limbo in which the *'repatriés'* found themselves in, unable to return to their businesses or their jobs in Senegal. Inspector General of Health Service Kermorgant in Paris again urged Governor General Ballay to adopt all indispensable measures to prevent the return of another yellow fever epidemic. He requested that sanitary councils from all localities urgently elaborate programs that could improve public hygiene, including garbage collection, water drainage, elimination of insalubrious housing, and crematory fours, among others, as well as the immediate construction and permanent maintenance of dissemination camps equipped with enough comfortable barracks to accommodate all European troops and colonists 'in order to subtract them from attacks from the disease.' There were reports that dissemination camps were successful in preserving European lives in Grand-Bassam (Côte d'Ivoire) and in Senegal, and the good results could have been even better with more comfort. So from Kermorgant's perspective, the best strategy to limit the ravages of the epidemic was 'to create a vacuum ahead of yellow fever,' meaning to separate the colonists from the *indigènes* through both isolation and repatriation.

Under pressure from the chambers of commerce, Minister of Colonies Decrais was prepared to go to the bottom of the matter and, having sought the opinion and the promise of support from the Pasteur Institute, he prepared a piece of legislation to be submitted to the Parliament requesting funds for the organization of a mission aimed at studying 'the nature of the infectious agent of typhus amaryl, its modes of transmission, and finally the preventive and curative treatment.' He also urged the colonial administrations on the West African coast 'to take all indispensable measures in order to ensure, in short terms, the implementation of the decree of March 31, 1897 (esp. articles 76–82, and 116) and to report on all the dispositions taken or projected in order to conform to different prescriptions of the decree.[71]

In the short run, faced with continuing reports of isolated cases of yellow fever in various localities of Senegal, one critical issue that continued to give headache to the authorities in Saint-Louis and in Paris was the decision to declare the epidemic over and to lift the travel ban to Senegal. Minister of colonies Decrais followed the recommendation of the inspector general of the health service that only the *'isolés'* (who were dispatched in isolation camps) and the military services with reduced staff be sent back to Senegal, and that the military units composed of European troops had to wait until further notice.[72] But Governor General Ballay, brushing aside the fears of re-emergence of the epidemic, instead favored the quick return of civil servants and soldiers, whose presence was seen as indispensable for the good conduct of affairs.[73] But the final word belonged to Paris.

The pattern of behavior of typhus amaril (another name for yellow fever) in 1900 seriously challenged the explanatory model followed until then, and the authorities had to account for the discrepancy. In his assessment of the government's response to the yellow fever epidemic, Kermorgant had attributed the official low mortality rates to two 'radical' measures taken at the beginning of the epidemic, that is, the dissemination of troops in isolation camps and the repatriation of European troops. From his perspective, yellow fever subsided simply for 'lack of food,' given that the Europeans on which it fed had left Senegal. He contended that, while previous epidemics 'generally departed with the return of the winds from the North' in January, this time the epidemic broke out earlier than usual and 'sporadic' cases continued to be reported well into the new season. The main reason he provided was the presence of 'new elements in the foyer,' that is, the personnel from the business firms who were 'precipitously' brought back to Senegal, some of whom becoming victims of the disease. He went on to explain that the 'persistence of the disease to strike newcomers is a proof of the great vitality of the infectious germ, and also a proof that one spark would suffice to reignite the foyer, and probably to lead to a true disaster.' This explanatory model attached to yellow fever some of the characteristics of miasma, including its kind of sleeping mode and sudden re-emergence following sanitation work or other ground disturbance that could 'reveal new germs.' Kermorgant expressed his fear of another outbreak of the 'disaster' (yellow fever) during the upcoming *hivernage* and strongly recommended that no new contingents of European troops be sent to Senegal in order to avoid 'providing food to the epidemic,' meaning the 'unseasoned young soldiers'.[74]

The death, on 18 January in Tivaouane, of Sibaud, a 17-year-old employee of Hess commercial firm and nephew of Député Ferrand, brought the yellow fever question to the floor of French parliament and asked questions to Minister of Colonies Decrais. Indeed, after learning about the death of his nephew, Ferrand requested explanations from the minister of colonies. Before Sibaud's trip to Senegal, the minister of colonies had informed him that the sanitary situation in Senegal was again satisfactory and presented no danger to his health. An internal investigation would later reveal that, in addition to the official declarations, passengers departing from Bordeaux for Senegal were also briefed about reports of new yellow fever cases in Senegal and that some passengers decided to postpone their trip on the basis of that information, and they received from the *Compagnie des Messageries Maritimes* full refund of their fares.[75] In response, Commissaire des Colonies Burrel, head of the Colonial Service in Bordeaux, confirmed having told passengers that in his view the only definitive signal of the end of the epidemic was the return of the cold season in January. 'No civil servant,' he added, 'has returned to Senegal without your orders.' Another piece of evidence was the postponement of the planned trip for recent graduates despite their eagerness to join their families. He went to a great length to explain that on 10 November, during his private conversations with civil servants, *négociants*, and other ship owners who were preparing to go back to Senegal, he made it clear to them that it was premature to return to Senegal, despite the fact that since October 26, in the absence of new cases of yellow fever, there was nothing to fear. He also circulated the same information on 16 November, after learning about a new yellow fever case in Tivaouane, even among the passengers who were ready to depart for Senegal and who decided to go anyway.[76]

The Chamber of Députés in Paris, in its session of February 15, 1901, questioned Minister of Colonies Albert Decrais about the 'worrisome' sanitary situation in Senegal. Député Stanislas Ferrand, after evoking the 'considerable ravages' caused by the 'disastrous' 1900 terrible yellow fever epidemic' and the repatriation to France of the Europeans, highlighted the lack of preparedness of the colonial authorities vis-à-vis the 'offensive return of the scourge' in terms of sanitary organization, health service, lazaretto, readiness of repatriation plans, as well as the 'acts of admirable civic courage performed by colonists, civilian administrators, and modest civil servants' under difficult circumstances. He deplored the granting of a medal of *Légion d'honneur* to Governor General Chaudié who, instead of adopting 'serious measures against the march of the scourge,' instead deserted his post at the beginning of the epidemic. Ferrand went so far as to qualify Chaudié's departure as an 'act of betrayal.' Having expressed his displeasure with disease preparedness and mismanagement, Ferrand blamed Decrais and the personnel in his ministry for the death, upon arrival, of the French personnel who went to Senegal after receiving assurances from government agents about the satisfactory sanitary situation in Senegal. In his view, Decrais either was ill informed by his services or he was not aware of the facts that were hidden from him, or he knew them and should have courageously let the country know; he stated that the official declaration of the end of the epidemic was a lie and that the death of his nephew Ernest Sibaud was not an isolated case. Ferrand also defined the situation in Senegal as a humanitarian question but also an economic one. According to his calculation, given that the epidemic crisis lasted more than eight months, the government had all the time to see and examine which measures could be successfully deployed to combat yellow fever, such as the construction of a lazaretto, the issuance of sanitary legislation, or the dispatch of physicians and drugs. From his perspective,

> The government had lacked vigilance handling the quarantines for we know, indeed, and many colleagues here know it better than me, that yel-

low fever was not endemic to Senegal. It originated in regions from the South of Senegal, in Brazil, in Cape Verde, brought by ships, by travelers, by merchandise. If quarantines were better managed, more rigorous, and if the sanitary service was better organized and more complete, we could have prevented the spread of this terrible scourge.[77]

In his critique of the administration of Senegal, Député Ferrand attributed the 'so worrisome frequencies' of yellow fever epidemic to shortage of medical personnel—a claim disputed by his colleague Hérisse—to complacency, and to a lack of well-organized sanitary service, especially in important commercial centers such as Tivaouane and Rufisque. He wanted from the minister of colonies explanations on the 'grave questions' raised and corrective measures in order to organize in Senegal 'a sanitary regime worth of France.'

In his response, Decrais revealed that he was planning to present before the National Assembly a project of law requesting funds to support a mission 'to conduct a scientific study of yellow fever where it exists at the endemic level,' in addition to another similar mission under way in Senegal—composed of physicians, engineers, and officers—that will propose 'all measures of hygiene, all sanitation works that will seem necessary.' He confessed being caught between contradictory demands from *negoçiants* from Bordeaux and Marseille, on the one hand, and from various circles, on the other. In a passionate tone, Decrais said the following: 'I am formally accused of having, by pusillanimity, by excessive prudence and protection, delayed the normalization of affairs and, by paralyzing the movement and life, inflicted a fatal blow to our colony.' He went on to explain his approach in face of these contradictory reproaches. 'Well,' he continued, 'I think I do not deserve either one of these contradictory reproaches,' and by following 'in all details, as I have done, the advice from a governor as sage, as prudent, as experienced as Dr. Ballay, *I believed to have conciliated the two interests in presence: the public health and commercial interest* (emphasis added).'[78]

This heated exchange between Minister of Colonies Decrais and Député Ferrand in the French parliament shows how what started as a health problem became an economic and political problem. The health crisis in Senegal, because of its long duration and its economic and political consequences, had become a great embarrassment for Paris. It underlined the difficulty to conciliate the interests of commerce and those of competing conceptions of public health. The association of yellow fever with the rainy season forced public health authorities to minimize the endemicity of yellow fever, using expressions, such as 'absolutely isolated and benign' cases, to explain the new reports of yellow fever that occurred during the dry season.[79] At the same time, Inspector General Kermorgant in Paris continued to recommend the most severe hygienic measures...in order to extinguish all those disseminated foyers (in the train stations along the railroad), which could be the point of departure of a new epidemic disease. The rigorous implementation of all measures of sanitary police by competent authorities constitutes the only means of defense that could protect the colony against a new explosion of (the) scourge.[80]

These measures were certainly inspired by the psychology of fear. A long-lasting solution was needed.

On March 7, Minister of Colonies Decrais presented his project of law requesting extraordinary credit to support the organization and maintenance of a scientific mission in order to study yellow fever in Brazil in 1902.[81] The story of the work of the scientific mission will be told in the next chapter.

## Conclusion

The evidence presented here shows that the power/knowledge alliance that had led to the imposition in 1884 of the automatic quarantine lasted almost a decade despite being seriously challenged by commercial interests. But the alliance slowly broke down under the combined pressure from the business community and supporters of the germ theory of disease causation. The factors of change also affected the internal cohesion of each group. The medical community remained largely attached to the contagionist theory since Louis Pasteur was a chemist and not a physician. If nothing else, the new theory only comforted them in their orthodoxy in that the germs simply replaced the miasma. But the re-emergence of yellow fever in 1900 and its persistence well beyond the rainy season raised pertinent questions about the internal validity of the explanatory model, weakened the power/knowledge alliance, and pushed the balance in favor of the new paradigm.

The political authorities in Paris, Saint-Louis, and the rest of French West Africa also had divergent opinions concerning the appropriate public health policy to deal with the threat of yellow fever. Although the sanitary legislation issued in Paris or in Saint-Louis provided clear guidance, its implementation depended on the local conditions and local interests at stake that needed to be accommodated.

The business community did not form a monolithic bloc either. It was also divided between contagionists—who were ready to sacrifice commercial interests in the name of one conception of public health, that is, the contagionist theory—and the followers of the germ theory that found in the new medical knowledge evidence that did not support the continuation of the quarantine system. At the end, it was the latter group that had the upper hand and convinced the political authorities in

Paris to support change. One needs to keep in mind that the debate about the quarantines can be understood in the framework of a larger debate that was going on in France and was animated by members of the Colonial Party, who were pushing for a policy of *mise en valeur* of the colonies. It was the commercial interests that convinced the minister of colonies to send a mission in inquiry to Senegal to find the deep roots of ill health and suggest remedies that did not support the use of quarantines in their current configuration. This story is the focus of the next chapter.

## Notes

1. ANFCAOM/FM/Sér. Géogr./Senegal/XI/34, excerpts from the Navy Health Superior Council meeting of Jan. 30, 1879.
2. ANFCAOM/Sér. Géogr./Senegal/XI/34, Rochard, general inspector of Health Service, to Navy and Colonies Minister, Sept. 20, 1878.
3. ANS/L13/12, governor's ordinance, May 30, 1873; early on, the domain was limited to the no man's land between Guet-Ndar and Ndar Toute, the place between the two garrisons, bridges, docks, and the routes to Cayor and Walo. ANFCAOM/ Sér. Géogr./Senegal/XI–28, Rapport au President de la Republique au sujet de la reorganization du regime sanitaire, July 29, 1884.
4. Later on, other ordinances prohibited the urban poor from drying animal skins and fish within the limits of the 'faubourgs,' or slum areas during the rainy season between 1 June and 15 December when the outbreaks of yellow fever generally occurred, and from throwing rubbish or even bathing in less than 150 meters on each side of the Servatius Drive and on Guet-Ndar beach. See ANS, Ordinance of May 23, 1885; ANS, BOS, municipal ordinances of Aug. 21, 1891, 368, and July 28, 1892, 345.
5. ANS, *Moniteur du Sénégal et Dépendances*, February 5, 1882, 23–24.
6. ANS, *Moniteur du Sénégal et Dépendances*, no. 1375 of 16 May 1882, 87.
7. ANS, 'Report of the Fourteenth of July celebration,' *Moniteur du Sénégal et Dépendances*, July 23, 1882, 155.
8. The law of 19 January, 7 March, and 13 April 1850 permitted the governor to choose no more than 9 and at least 5 members, including a physician, an architect or an artist, and a representative of a charity group, if applicable. The commission was presided over by the mayor or his deputy; it renewed or reappointed a third of its members every two years (art. 2). ANS, 'Decree of September 15, 1882, *Moniteur du Sénégal et Dépendances*, October 1882, 182–83.
9. ANS, Moniteur du Sénégal et Dépendances, October 1882, 182.
10. ANS, MSD, ordinance of May 10, 1882; ordinance of Jan. 19, 1889; Mayor A. de Bourgmeister's ordinance of Febr. 25, 1889. The administration of sanitary police was then exercised by the newly created post of Interior Director.
11. Maurice Nielly, Eléments de pathologie exotique (Paris: Delahaye et Lecrosnier, 1881).
12. On July 18, 1882, medical officials in Dakar imposed 21-day quarantines on ships arriving from the south between Pointe Sangomar and Gabon with an unclean bill of health, and at least 7-day quarantines of observation on ships without a bill of health. Governor A Vallon

still hoped the measure 'would facilitate the operations of commerce while protecting the colony, as much as possible, from an invasion of yellow fever or any other infectious disease.' Actually, quarantine measures constituted a major obstacle to commercial activities and a loss of revenues, as we will see later in this chapter.

13. ANS, *Moniteur du Sénégal et Dépendances*, December 5, 1882, 225.
14. ANS/H22/Senegal, Ordinance of June 20, 1883 prescribing the measures of sanitary police for Senegal and Dependencies.
15. ANFCAOM/FM/Sér. Géogr./Senegal/XI-28, Commission's report to governor, Dec. 29, 1883.
16. ANFCAOM/Sér. Senegal/XI-28, Ordinance of governor of Senegal and Dependencies, June 19, 1884.
17. ANS/H2/Senegal, decree on the reorganization of the sanitary service in Senegal, 29 August 1884. This decree was similar to the decree signed on November 1883, which protected France against the suspected ships coming from America.
18. ANFCAOM/FM/SEN/XI/50A, Sanitary Commission meeting minutes, June 6, 1899.
19. ANS/H37/AOF/58, Merchants and traders of Sierra Leone to Mr. V. Bareste, Sept. 12, 1885.
20. ANS/H2/Senegal, report about quarantines by chief medical officer, May 3, 1889.
21. David Arnold, *Colonizing the Body: State Medicine and Epidemic Disease in Nineteenth-Century India* (Berkeley and Los Angeles: University of California Press, 1993), 189, 191–92, 194–95, 197.
22. ANS/H2/Senegal, governor to interior director, no. 139 of May 27, 1889; ANS/H2/Senegal, E. Etienne, undersecretary of state for colonies to governor, no. 778 of June 29, 1889.
23. ANS/H4/Senegal, minutes of the Council of Public Hygiene and Salubrity, May 31, 1890.
24. ANFSOM, Série géogr. Senegal XI d. 31.
25. ANS/H4/Senegal, minutes of the Sanitary Commission meeting, 25 Aug. 1892.
26. ANS/H4/Senegal, minutes of the Sanitary Commission, Sept. 1, 1892.
27. ANS/H46, decree of 29 Dec. 1893 modifying the decree of 29 Aug. 1884.
28. ANS, *Bulletin Officiel du Sénégal*, 1897, 343.
29. ANS/H2 (A.O.F.), law related to the sanitary police, title II, art. 7–14, March 3, 1822.
30. Ibid., art. 10.
31. ANS/H4/Senegal, Chamber of Commerce to Colonial Commission, Sept. 11, 1898.
32. ANS/H24/Senegal, minutes of the Sanitary Commission meeting, 17 May 1899.
33. ANS/H25/Senegal, Chief medical officer to governor general, no. 610 of June 2, 1899.
34. ANFCAOM/FM/SEN/XI/50/A, Deliberations of *Conseil Privé* meeting, no. 12 of June 8, 1899.
35. ANFCAOM/FM/SEN/XI/50A, Sanitary Commission meeting records, June 6, 1899; report prepared by physician Lafage, Chief Medical Officer, and read at the meeting, suggesting more rigorous measures.
36. ANS/H24/Senegal, the Sanitary Commission meeting records, June 6, 1899.
37. The *Conseil Privé* included Govenor General Chaudié (president), Bergès (secretary general of government), Cnapelynck (general prosecutor, head of judiciary service), Bunel (commissar of colonies, head of administrative service), Chief Medical Officer Lafage, R. Martin (member), Gros (member suppéant), and Sasias (archivist secretary). Invited guests for the session held on 8 June 1899 were Colonel Pujol (Superior Commander of Troops), Germain

d'Erneville (Chamber of Commerce president, Saint-Louis), Theodore Carpot (deputy mayor filling in for the mayor), Fr. Marsat (mayor of Dakar), Gabard (mayor of Rufisque), and Le Bègue de Germiny (mayor of Goree).

38. ANFCAOM/FM/SEN/XI/50A, The *Conseil Privé* meeting records, June 8, 1899.
39. ANFCAOM/FM/SEN/XI/50A, Governor General Chaudié to minister of colonies, no. 1180 of June 15, 1899.
40. Ibid.
41. ANFCAOM/FM/SEN/XI/50A, Note from inspector general of Health Service in Paris to Director of Africa Desk (1st Bureau), July 12, 1899.
42. ANS/H42/104, 38 European Residents of Tivaouane to Administrator of Tivaouane, May 20, 1900; ANS/H108, Administrator Decressac-Villagrand to governor general, 4 July 1900.
43. ANFCAOM/FM/SEN/XI/50A, 'Population Tivaouane to minister of colonies, August 15, 1900.
44. ANFCAOM/FM/SEN/XI/50A, Note pour la 1ere Direction le Bureau, by Kermorgant, August 22, 1900; see also Decrais to governor, August 23, 1900.
45. ANFCAOM/FM/SEN/XI/50A, Note pour l'inspection générale du service de santé, by Kermorgant, Aug. 30, 1900.
46. ANFCAOM/FM/SEN/XI/50A, Lans et Henri, Aug. 26, 1900.
47. ANS/H 41/AOF/75, Société le Syndicat du Soudan Français, September 7, 1900.
48. ANFCAOM/FM/SEN/XI/50A, Minister of colonies to minister of navy, Sept. 1, 1900.
49. ANFCAOM/FM/SEN/XI/50A, General Combes to governor general, Sept. 13, 1900.
50. ANFCAOM/FM/SEN/XI/50A, Governor General Ballay to minister of colonies, Sept. 18, 1900.
51. ANFCAOM/FM/SEN/XI/50A, Ballay to minister of colonies, Sept. 15, 1900.
52. ANS/H25/Senegal, minutes of the Municipal Council presided over by Mayor Descemet, 30 July 1900.
53. ANFCAOM/FM/SEN/XI/50A, Signatories to minister of colonies, July 22, 1900; the letter was signed by Gouzy, Dordain, Fabbry, Gersan, and Heldt Lébire.
54. After investigations, Chaudié was appointed inspector general in the Ministry of Colonies on October 18, 1901 with effects on January 1, 1902. Chaudié requested to be sent back to Senegal during a future epidemic either as an inspector general in mission or in another capacity in order to put to rest all accusations against him. He remained convinced that only a voluntary presence in Senegal during an epidemic would prove his detractors wrong (see ANFCAOM/FM/SEN/XI/50A/14, Chaudie to minister, Oct. 9, 1901).
55. ANS, *Bulletin Administratif, 1900*, decrees of 23 and 26 July.
56. ANFCAOM/FM/SEN/XI/50A, E. Guérard, Secretary of Administrative Council, to minister of colonies, Aug. 11, 1900.
57. ANFCAOM/FM/SEN/XI/50A, Ballay to minister of colonies, Dec. 2, 1900; Decrais to Sulpicy, Dec. 3, 1900. Sulpicy inquired about traveling to Senegal.
58. ANFCAOM/
59. Annick Guénel, "The Creation of the First Overseas Pasteur Institute, or the Beginning of Albert Calmette's Pastorian Career," *Medical History* 43 (1999), 1–2. The bacteriological laboratory of Saint-Louis was built in 1897; see C. Mathis, *L'Oeuvre des Pastoriens en Afrique Noire*, (Paris: PUF, 1946); Anne Marie Moulin, "Patriarchal Science: The Network of Overseas Pasteur Institutes," in Patrick Petitjean, Catherine Jami, and Anne Marie Moulin (eds.),

Science and Empires: Historical Studies About Scientific Development and European Expansion (Dordrecht; Boston: Kluwer Academic Publishers, 1992), 307.
60. Marc Renneville, "Politiques de l'hygiène de l'AFAS (1872–1914), in Patrice Bourdelais (dir.), *Les Hygiénistes: enjeux, modèles et pratiques* (Paris: Éditions Belin), 2001, 86–87, 90.
61. Ibid., 85, 90.
62. Walter Reed, "Recent Researches Concerning the Etiology, Propagation, and Prevention of Yellow Fever, by the United States Army Commission," *The Journal of Hygiene* 2.2 (April 1, 1902), 104–107, 111–15, 118–19.
63. ANFSOM/FM/SEN/XI/50A, Administrator-Director to Minister of Colonies, Nov. 3, 1900.
64. ANFCAOM/FM/SEN/XI/50A, Administrator-director of the Compagnie Française de l'Afrique Occidentale to minister of colonies, Nov. 3, 1900.
65. ANFSOM/SEN/XI/50/A, Administrator-Director of the Compagnie Française de l'Afrique Occidentale to the minister of colonies, Nov. 3, 1900.
66. ANFCAOM/FM/SEN/XI/50A, Governor General Ballay to minister of colonies, Dec. 23, 1900.
67. ANFCAOM/FM/SEN/XI/50A, Ballay to minister of colonies, Jan. 19, 1901.
68. ANFCAOM/FM/SEN/XI/50A, Governor General Ballay to minister of colonies, Nov. 23, 1900.
69. ANFCAOM/FM/SEN/XI/50A, ibid., Dec. 23, 1900.
70. ANFCAOM/FM/SEN/XI/50A, E. de Traz, President of Administrative Council of the Compagnie de Chemin de Fer de Dakar a Saint-Louis, to minister of colonies, Nov. 14, 1900.
71. ANFCAOM/FM/SEN/XI/50A, Note pour la 1e Direction 1e Bureau, by Kermorgant, Nov. 16, 1900.
72. ANFCAOM/FM/SEN/XI/50A, B. Drouhez, Note pour l'Inspecteur Général du Service de Santé, Jan. 10, 1901.
73. Idem., Jan. 19, 1901.
74. ANFCAOM/FM/SEN/XI/50A, Kermorgant, report to minister, Jan. 7, 1901. In fact, the only Europeans who were sent to Senegal since the end of the epidemic were 5 officers and 8 soldiers (Joint Chief of Staff), 13 officers and 16 junior officers and corporals (Tiraileurs Regiment), and one Senegalese officer; see Note sur l'epidemie de fievre jaune au Senegal, by Drouker?, Feb. 1901.
75. ANFCAOM/FM/SEN/XI/50A, Albert Decrais to Head of Colonial Service in Bordeaux, February 12, 1901; see also Ballay to minister of colonies, Jan. 19, 1901.
76. ANFCAOM/FM/SEN/XI/50A, Burrel, head of Colonial Service in Bordeaux, to minister of colonies, Febr. 13, 1901.
77. ANFCAOM/BIB, Chamber des Députés, Débats parlementaires, Session of Febr., 15, 1901, 439.
78. ANFCAOM/BIB, ibid., 440.
79. ANFCAOM/FM/SEN/XI/50A, Ballay to minister of colonies, March 6, 1901.
80. ANFCAOM/FM/SEN/XI/50A, Note to le Direction le Bureau, by Kermorgant, March 28, 1901.
81. J.O. of 8 March, 633.

Plate 1. The Faidherbe Bridge next to the old bridge. Postcard

*Plate 2. A street in Saint-Louis*

## 156 | COLONIAL PATHOLOGIES, ENVIRONMENT, AND WESTERN MEDICINE

Plate 3. A street in Saint-Louis

Plate 4. A slum area at the south end of the city island. Postcard

158 | COLONIAL PATHOLOGIES, ENVIRONMENT, AND WESTERN MEDICINE

Plate 5. A view of Guet-Ndar slum. Postcard

ILLUSTRATIONS | 159

Plate 6. The small arm of the Senegal River

160 | COLONIAL PATHOLOGIES, ENVIRONMENT, AND WESTERN MEDICINE

Plate 7. A mule-drawn cart in Ndar–Toute quarter

*Plate 8. The Grotte de Lourdes in Sor*

*Plate 9. The monument to the victims of the 1878 yellow fever epidemic*

ILLUSTRATIONS | 163

Plate 10. *Water tank in Sor. Postcard*

CHAPTER FIVE

# The Scientific Missions to Senegal and Brazil and the New Paradigm, 1901–1912

The 1900 yellow fever epidemic and the panic and chaos it provoked in Senegal forced the French government to seek a permanent solution to the permanent threat of yellow fever epidemics. To this effect the authorities funded two missions: one to Senegal in February–March 1901 and another to Rio de Janeiro in November 1901. Their stories will be told here.

## The Grall-Marchoux-Jacquerez Sanitary Mission to Senegal, February–March 1901

### Background to the Mission of Inquiry

The idea of sending a team of doctors and engineers to Senegal to identify the causes of yellow fever and to suggest solutions to the health crisis was first put forward by Jacques D'Urville, a journalist, in *L'Echo de Paris* of 5 September 1900, during the yellow fever epidemic. In an article titled 'La Fièvre Jaune au Sénégal,' he vehemently attacked Governor General E. Chaudié and his collaborators for their cowardly act of abandoning their posts in the middle of an epidemic crisis, thus paralyzing of the administration and commerce. He pointed out that

> the ravages caused by yellow fever in Senegal are already considerable. It is urgent to send there a special medical mission charged to study the disease

and to determine the curative and preventive methods to apply....physicians in Senegal, from what is alleged, hesitate to decide about the means to use in order to combat the disaster. In Saint-Louis, patients are treated with iced water; in Dakar, they make patients sweat abundantly....M. Chaudié did not hesitate to affirm that any white person suffering from yellow fever is in advance condemned.[1]

Jacques D'Urville identified what he believed was the root cause of the health crisis in Senegal, that is, the inability of doctors to correctly identify the infectious agent and to determine an effective and uniform treatment for yellow fever; in other words, the crisis had to do with the imperfect knowledge about its epidemiology and disease etiology. Doctors were well aware of their own shortcomings. Indeed, in the aftermath of the 1900 yellow fever epidemic, Dr. Charles Carpot, Director of the Civilian Hospital in Saint-Louis, explained that there was no consensus concerning the nature of the 'yellow fever microbe': some medical officials claimed that the *Peronospora Lutoea* was its etiological agent; others contended that it was the *Cryptococcus Xantogenicus* instead, while another group of health officers believed that yellow fever was caused by the transformation of human "leucomains" into "*ptomains*," as a result of poor hygiene and diet, moral preoccupations, physical fatigues, and overcrowding. Dr. Carpot found the last theory less plausible because yellow fever had the predilection of producing its victims preferably among strong individuals and the newcomers. He also rejected the 'seaweed-generated miasmas/microbes/mushrooms' as the causes of yellow fever. He repeated the received wisdom that the most susceptible groups were the adult Europeans and the African children (from the Creoles' and the *indigènes*' families), and that immunity was acquired after surviving the first infection, not through acclimatization.[2]

There were also commercial considerations, especially the merchants' energetic opposition to the quarantines that were not justified on scientific grounds but were detrimental to trade, as seen in chapter 4. Merchants were particularly concerned with the increasing marginalization of Saint-Louis, a city with an underdeveloped port not easily accessible for ships because of the bar. In addition, the suggestion to send a special medical mission to Senegal found supporters within the Ministry of Colonies.

## The Investigation

On February 2, 1901 the Sanitary Mission received specific ministerial instructions concerning the aim of the mission: to investigate the sanitary situation of the colony and to recommend an appropriate sanitation program. After obtaining the ministerial instructions, the Sanitary Mission traveled to Senegal to start its investigation. It included Dr. D. Grall, *médecin inspecteur* and president of the Sanitary

Mission, Dr. Emile Marchoux, member, engineer Jacquerez, inspector at the Public Works Service, and Captain Rambaud, from artillery. Dr. Grall and Dr. Marchoux knew Senegal from having worked there in 1890. Dr. Marchoux was again in Senegal in 1897 to open the bacteriological laboratory in Saint-Louis. They understood their mission to be a consultative one.

In Senegal the members of the Sanitary Mission visited the garrisons and hospitals in Saint-Louis, Dakar, and Gorée. They also investigated the sanitary situation in these three cities as well as in the towns located along the railroad Saint-Louis-Dakar, including Thiès, Rufisque, Tivaouane, and Louga, where there was a substantial French presence. They spoke with the French traders and the administrative authorities about their understanding of the health problems and the solutions to resolve them. They did not seek the opinions of the *indigènes* and their leaders, such as slums and village chiefs, because they shared with the French colonial and medical officials the basic assumption that the *indigènes* had developed immunity against yellow fever, that they were carriers of infectious agents, and that they were used to taking their water directly from the rivers without great danger for their health. Their main preoccupation was with the health of the Europeans. Their main assumption was that hygiene was one of the most important attributes of the Civilizing Mission and that the colonial rule over vast intertropical regions would be impossible without hygiene.

On March 4, 1901, Dr. Grall attended a meeting organized by the Hygiene Municipal Committee. The participants discussed the recommendations made by the Sanitary Mission, and the role the city would play in terms of improving city sanitation, and enforcing the existing sanitary legislation. He also informed the Ministry of Colonies about the sanitation program to be executed for the city of Saint-Louis.[3] He attended a meeting held by the Hygiene Colonial Committee at the *Hôtel de Ville*. The discussion evolved around the need to abandon the outmoded system of automatic annual quarantine between 1 June and 15 December (art. 113 of the decree of 29–8–1884) in favor of sanitary measures compatible with the new medical knowledge, such as draining the swamps in the northern part of the city, in Sor, and on Bop N'Kior Island.[4] After discussing their findings with the administration and municipal authorities, the Sanitary Mission brought its investigation to an end and left for France. The mission wrote its final report, including specific recommendations.

## The Final Report

The Sanitary Mission based its conclusions and recommendations on the observations the experts made on the ground, the meetings they had with various authorities (the governor, the heads of local services, the presidents of elective assemblies, and the *habitants notables*, that is, *métis*), and on their own opinions. The main gray area remained the etiology and epidemiology of yellow fever.

The final report dealt with three main health sectors: military hygiene (Dr. D. Grall), urban hygiene (Dr. Emile Marchoux), and technical issues such as drinking water, sewers, rubbish collection, and areas of standing water such as swamps, bogs, and moats (engineer Jacquerez). The experts first provided a general overview of the ravages caused by the diseases, especially yellow fever and malaria, before presenting detailed recommendations on improving the sanitary situation in the garrisons, the city, and the quality of public health infrastructure, on better medical personnel management, and on the provision of amenities to the city. Overall, the problems were the same in these agglomerations but the emphasis here will be on the aspects of the reports related to Saint-Louis.[5] I shall follow the order of the reports as presented by the experts.

## MILITARY HYGIENE

Dr. Grall's main argument was that the high mortality rate among the European troops from the Navy Infantry, Third Company, housed in Rognat garrisons, and among the medical staff in the Colonial Hospital resulted from insanitary conditions of the rooms where the victims had lived or worked. The officers who came down with yellow fever were in 'compulsory relations' with the civilian population. In the hospital the cohabitation between the medical staff, the ordinary sick, and the patients with contagious diseases contributed to the spread of yellow fever among the Sisters of Saint-Joseph de Cluny. Dr. Grall explained that there was a correlation between the rapidity and the dangerous nature of the contagion and the location of the Rognat garrisons 'in the middle of the agglomeration, without a protection zone and without separation.' In contrast, the *Spahis* and the *Tirailleurs* garrisons that were located outside the agglomeration—respectively in the northern part of Saint-Louis Island and north of Ndar-Toute—experienced fewer cases of yellow fever. The evacuation and isolation of troops to Louga, a train station located between Saint-Louis and Dakar, saved many lives because the disseminated troops did not take their (contaminated) clothes with them, and they were repatriated to France fairly quickly.

Another contributing factor to mortality, according to Dr. Grall, was related to the treatment of yellow fever patients. Indeed, at the beginning of the epidemic, suspected cases and 'abortive' cases were neglected, and an effective treatment started only on the third or fourth day after the onset of the epidemic. He also observed that mortality among the Europeans depended on the duration of the stay in the colony; the Catholic Sisters who had 'a prolonged stay in the colonies' and had acquired immunity, suffered less from the ravages of yellow fever than the young troops still at the beginning of their colonial service.[6]

Dr. Grall's theory of yellow fever, which combined miasmatic, contagionist, and germ ideas, deserves some attention here. He reiterated some of the postulates of the miasmatic theory. For example, he described yellow fever as 'a disease of the interior of the house' that spreads by living either in contaminated and filthy rooms for a long time or in close proximity with them, but not through occasional contact with the patients. He also stated that 'the patient, his furniture, clothes, and bed sheets conserve and spread the disease. In his view, the disease agency, called the *contage*—another term for miasma—if not properly destroyed by an 'effective and sufficient' disinfection, could revive and regain potency without outside help. Otherwise, he contended, it would not be possible to explain the epidemic outbreak foyers inside the Catholic Sisters' convents in Dakar and Saint-Louis, the nurses' dormitories, and the Gorée and Rognat garrisons during the 1900 yellow fever epidemic.

Concerning the mode of transmission of yellow fever, he put forward a 'tentative conclusion' subject to additional verification. From his perspective, the agent of transmission of yellow fever was

> one of the parasites that crowd the house. These insects attack people and inoculate them with the germ obtained from the patient's blood or from his bodily defilements. It could be a mosquito in certain cases; it must be more frequently one of the varieties of ticks that are so numerous in tropical regions.[7]

The enumeration of these 'creatures' that populated the 'malfeasant world,' including the parasites, insects, mosquitos and ticks, shows that Dr. Grall was trying to understand the etiology of yellow fever in the light of the recent discoveries on the etiology of malaria (made by Ross in 1898) and plague (Simond in 1898). He argued that mosquitos played a role in the etiology of yellow fever 'in certain cases,' not in the majority of cases. He proposed, instead, what can be termed a 'tick theory,' to which he attributed a greater analytical power than other competing theories. For evidence in support of his theory, he pointed to the slow pace and scopes of the spread of yellow fever, and the preponderant action that individual and domestic cleanliness and comfort had on its development. Dr. Grall suggested that only his 'tick theory' could cope with unexplained particular 'facts,' such as the spread of the disease by individuals living in contaminated houses who were carriers but did not exhibit the symptoms of the disease.

Dr. Grall was well aware of the controversial character of parts of his etiological and epidemiological theory of yellow fever, but he contended that the general theoretical framework was sufficiently well established to provide the foundation for 'the rules of a reasoned hygiene for the troops' in regions periodically visited by yellow fever. Dr. Grall was certainly struggling with the concepts of germ, vehicle, and host, and he had problems distinguishing between the mode of transmission and diffusion, like many other contemporary physicians.[8]

In his conclusion with respect to military hygiene, Dr. Grall underlined the importance of hygiene and of a protection zone in maintaining health. He recommended the isolation of the sick (their admission into the hospital) as soon as the first signs and symptoms of the disease appeared, instead of keeping them in their rooms or in the infirmary for three or four days before providing them with an efficient treatment. To provide the troops with such protection required that a certain number of conditions be fulfilled, such as the transfer of the Colonial Hospital to Pointe Nord near the mosque after draining the swamp, the provision of the hospital with large and clean rooms, enough ventilation and light, and the building of a special hospital or a special room for persons with infectious diseases on the Langue de Barbarie between Guet-Ndar and the Muslim cemetery.

Dr. Grall knew that the medical infrastructure alone would be useless if there were no permanent and motivated personnel. He recommended the recruitment and training of specialized personnel, who could work in the colony for a long period in order to get acclimated to the local conditions. Before 1900 the majority of the victims were young people who had recently migrated to the colony and not long-term residents like the medical personnel. In order to stay healthy, the personnel would be required to observe strict rules of hygiene, including washing and disinfecting the medical personnel's and patients' clothes and bedclothes. This required the installation of administrative laundries.

Dr. Grall paid a special attention to the living conditions of the soldiers. He had observed that soldiers were not sufficiently protected against the arthropod vectors of yellow fever, malaria and other diseases such as mosquitoes, gnats, and ticks that were found in great numbers in the rooms. The best way to protect soldiers would be to prevent them from sleeping on the first floor and to use mosquito-nets around the bed. Another recommendation related to the cleanliness of the rooms and the premises of the garrisons. This included the renovation and enlargement of the rooms, the elimination of the workshops, stores, and other used furniture, which were annexed to the garrisons and were difficult to clean. The Rognat garrison, considered a 'foyer of infections,' was to be discarded as housing for the European troops. In order to avoid overcrowding conditions, the administration had to build new garrisons on the periphery of the city; the infantry garrisons would be built either across the street from the future hospital in Pointe Nord or in Guet-Ndar.

Convinced that the *indigènes* were potential carriers of diseases, Dr. Grall recommended a form of 'sanitary segregation' in order to prevent the European troops from interacting with the African troops (that is, *Tirailleurs, Spahis,* and *Conducteurs*). The medical view was that African troops were susceptible of contracting the disease in the slums, where they lived, and to transmit it through their clothes, supplies, and vermin, and through shared housing. A limited interaction would be authorized with 'an indispensable fraction of their (European) cadres' necessary for the daily running of the service. Another possibility would be to put them in a 'military village,' like in Cochinchine, in the periphery of the city beyond Ndar-Toute. This case shows that, in the aftermath of the yellow fever epidemic, the 'sanitation syndrome' had reached its climax.

## URBAN HYGIENE

Dr. Marchoux's investigation on urban hygiene focused on sanitary conditions in the cities, and the immediate as well as long-term changes that the situation required. His approach to the health crisis in Senegal was guided by two assumptions: first, there was a correlation between the outbreaks of epidemics and public and private hygiene; second, hygiene was the *sine qua non* condition of the 'true conquest' of the inter-tropical zone. Like Dr. Grall, and based on the knowledge that malaria was transmitted by mosquitos, he thought it possible that yellow fever also would be transmitted by the same agents. Wherever he went, he paid close attention to the sanitary conditions.

Dr. Marchoux found that Saint-Louis had no specialized service for the collection and disposal of waste. City and slum residents threw their waste in the Senegal River. The river's edge was littered with abandoned canoes, stones, animal wastes, fish bones, kitchen garbage, human excrements, and other petrifying matters. The small branch of the river was transformed into a cesspool. Dr. Marchoux observed that these insanitary habits continued unopposed by the authorities. 'Real ponds' surrounded water fountains installed on the docks. At the extreme north and south of the city island, he visited the urban poor who lived in overcrowded dwellings that they shared with domestic animals.

Dr. Marchoux described Guet-Ndar as an overcrowded slum with houses 'pressed against one another.' The dwellings, although cool during the dry and cold season, were unfit for occupation and were easily destroyed by fires. The fishermen 'dumped waste and kitchen garbage everywhere,' including places where children played and adults spent their days. Fishermen left fish to dry on the roofs of their dwellings, thus spreading noxious odors. They shared their dwellings with sheep, goats, and sometime cows; and, as a result, the soil inside the dwellings was covered with organic wastes. Women gave birth 'assisted by midwives with dirty hands' in these insanitary conditions. The marketplace, located in Guet-Ndar at the entrance of the Serviatus Bridge, and the two nearby parks where animals were sold, greatly contributed to the city's environmental crisis. Dr. Marchoux concluded that poor sanitation in Guet-Ndar was responsible for the 'frightening ravages' of tetanus among the newly born children as well as pneumonia among the fishermen. In Sor he observed that part of the island was flooded during the rainy season, and it constituted the breeding ground for mosquitos.

At the end of his investigation, Dr. Marchoux recommended the promotion of cleanliness in the cities and dwellings, the importance of the provision of clean drinking water using the methods of sterilization and ozonisation as a prophylactic measure against typhoid fever and dysentery, and the draining of areas of standing water. The swamps classified as 'most dangerous' were at six locations: 1) the northern part of Saint-Louis; 2) the zone comprised between the railroad, the street leading to the train station, and the river; 3) the pond in Bouetville (Sor), including the ground across the street from the church; 4) the ground beyond the train station for merchandise on the right side of the railroad; 5) the extremity of the island of Bop-N'Kior; and 6) the low grounds between cemetery no. 2 and cemetery no. 3 in Sor. These swamps had to be eliminated. The embankment would involve 1,055,000 m$^3$.

Dr. Marchoux made suggestions concerning the construction of a slaughterhouse in the city, and the elimination of the discarded cemetery at the entrance of the Faidherbe Bridge in Sor.[9] He also recommended the completion of the construction of the docks around the island, and the construction of a toilet system equipped with flushing on the small arm of the Senegal River. Another recommendation concerned the forced removal of 'primitive housing' (and of the poor) from the island and their transfer to Sor (Bouetville) once the draining of the swamps was completed.

## 'TECHNICAL' ISSUES

Engineer Jacquerez examined 'technical' issues, such as water supply, the sewer system, and garbage collection. Concerning water supply, he described past attempts to bring fresh drinking water to Saint-Louis, the unresolved problems that the city still faced in the 1890s, and the alternative solutions being implemented. With respect to the problem of waste disposal, Jacquerez recommended the construction of long wharfs south of Saint-Louis, and along the small arm of the Senegal River equipped with public toilets at their extremity, where the "*marseillaises*" (vases used as toilets) containing human waste could be emptied. A more sophisticated system would be built later.

Jacquerez observed that garbage and kitchen wastes littered the city streets, even closed some streets leading to the river, such as Saint-Jean, Flamand, Saint-Paul, Mage et Baron Roger, and caused disease. He recommended the rebuilding and cleaning of city streets, the organization of a better system of garbage and refuse disposal, and an effective implementation of the decree issued by the Mayor on January 18, 1882, on cleanliness. A constant surveillance system by a special brigade of police agents would ensure the strict respect of sanitation ordinances. The sanitation program would be incomplete without the construction of docks around the city, and the urgent drainage of the existing bodies of standing water, especially the huge pond located between the prison and the artillery garrison which constituted 'a permanent cause of insalubrity.' The total cost of the sanitation program in Saint-Louis was estimated at 4,900,000 francs, without taking into account the cost of improving the water system.[10]

## Follow-up to the Sanitary Mission

In a letter dated March 30, 1901, Dr. Grall proposed special measures to be taken for the 1901 *hivernage*, including the separation of African troops, 'active carriers of the epidemic,' from the European troops; and the dispatch of the 2 officers and 60 troops in a camp located at Pointe aux Chameaux and of the essential cadres in a camp in Ndar-Toute.[11] The measures reflected the 'sanitation syndrome' that characterized the colonial mentality. But the *hivernage* came without any serious threat of yellow fever.

On June 1, 1901, Dr. Grall went back to Saint-Louis to attend a special meeting of the Special Commission, presided over by Governor General N. Ballay, charged to examine the recommendations of the Sanitary Mission; his presence was necessary to provide additional information to the Special Commission.[12]

The interpretation of the main findings of the Sanitary Mission led to tension between the administration and the municipality. Indeed, in their analysis of the factors which influenced the re-emergence of infectious diseases in the city, especially yellow fever, the experts of the Sanitary Mission had underlined the inability of the bureaucracy to enforce the existing public sanitary measures. From their perspective, yellow fever surprised everyone because

> The sanitary legislation had become loose, if not in its leading ideas, at least in its interpretation and its application; the sanitary equipment was neither completed nor improved according to the directives provided by the decree of 1897—the personnel in charge of the implementation lacked experience and was understaffed; the lessons from the past were forgotten.[13]

Dr. Grall blamed the physicians for doing a sloppy job in the diagnosis of yellow fever, the isolation of the sick, and the supervision of the disinfection teams. However, he was particularly harsh on the municipal authorities, who, because of their 'apathy and carelessness, facilitated the spread [of the disease] and created dangers for the future.'[14] Dr. Grall simply repeated what the public health officials had expressed in other circumstances, that is, the state of city sanitation was the responsibility of the municipality; the latter had to take the blame for the dirty and unsanitary city, and the spread of yellow fever epidemic.

The municipality had a different understanding of the issue, however. In a letter, dated September 25, 1901, to the governor general and president of the Special Commission examining the conclusions of the Sanitary Mission, Theodore Carpot, president of the *Conseil Général* and deputy-mayor (and brother of Dr. Charles Carpot), vehemently protested against the allegations of negligence made by the members of the commission. He argued that the public health measures were enforced until the day Dr. Lecorre, Chief Medical Officer, president of the Sanitary Commission and of the Local Hygiene Committee, provided reasons for ending the enforcement of the public health measures. Theodore Carpot reminded the governor general that the power of the city authorities was not boundless, and that the sanitary laws themselves were flawed. He concluded that the true origin of yellow fever and its mode of transmission had to be found somewhere else.[15] These reactions reflected the ongoing conflict between the municipal authorities and the administration, which reflected competing agendas and ideologies, and other unresolved problems between the two groups. The French authorities blamed the municipal authorities (Creoles) for protecting the interests of the local population (election politics) and jeopardizing the city's health. After the initial phase of accusations, counter-accusations and denunciation of various officials, the administration and the municipality prepared their (separate) responses to the recommendations of the Sanitary Mission. Given that the Sanitary Mission in Senegal did not resolve the key questions of the identification of the agent of yellow fever and its mode of transmission, and in the light of fresh discoveries being made in Cuba, the French government had to find another solution.

## The Simond-Marchoux-Selimbeni Mission in Brazil, November 1901–May 1905

The discovery made by Walter Reed and the American scientific mission in Cuba that a mosquito called *Stegomyia fasciata* played a role in the transmission of yellow fever was received with skepticism by part of the scientific community. There was a need to confirm his findings, thus leading to the multiplication of scientific missions to South America, such as John Guiteras to Havana; Emilio Ribas and Adolfo Lutz to São Paulo; and Parker, Beyer and Pothier to Vera Cruz. Reed's discovery, combined with the news that the sanitation campaign rid Cuba of yellow fever, prompted the French government to send a scientific mission to Rio de Janeiro to investigate new approaches to yellow fever control. The leading investigator was physician Paul-Louis Simond, who in June 1898 had discovered the role of the flea in the transmission of plague. His discovery was critically important in combating plague at the end of the nineteenth century. He was assisted by Emile

Marchoux and Alexandre Salimbeni. All three were members of the Institut Pasteur in Paris, whose director, Emile Roux, had placed his full confidence in them. Between November 1901 and May 1905, at the São Sebastiano Hospital, they conducted experiments with the *Stegomyia fasciata* female mosquitos and on human subjects, who were exposed to the bite of infected mosquitos, in order to understand the mode of transmission of yellow fever. Their experimentation also led to the making of the first yellow fever vaccine using the convalescents' serum.

In their first publication in 1903, they helped define the objective criteria for the identification of yellow fever cases, the matter that had confused physicians who often easily mistook malaria cases for yellow fever. One of their main findings from the experimentation undertaken in early 1905 was that the agent of yellow fever (which still remained to be identified) could be transmitted from an infected female *Stegomyia* to its eggs and larvae. Their findings led to the adoption of new sanitary rules to combat the spread of yellow fever by destroying not only the adult mosquitoes but also the eggs and larvae. From their epidemiological study of yellow fever in Rio since 1890, they discovered that, unlike the received wisdom, children did not have a natural immunity to yellow fever that they would lose as they grew older. Instead, they found that the vulnerability to yellow fever was the same at all ages but that often children developed a benign form of yellow fever that went unnoticed but that gave them immunity. Simond and Marchoux also established that the adult natives did not have natural immunity to yellow fever either; instead they acquired a relative immunity after the first attack of the disease during childhood that would be maintained by reinfection and would prevent severe forms of disease.

By the end of 1903, Simond and Marchoux made public the preliminary conclusions of their experimentations and the main principles of prophylaxis for practical use, including the systematic destruction of mosquitoes and larvae, and the protection of the sick from the *Stegomyia* mosquito bites through the use of a screened metallic cage. The adoption of these new sanitary rules contributed to the success of the 'anti-yellow fever campaign' in the city of Rio led by physician Oswald Cruz, a young Brazilian bacteriologist who was recently appointed federal Chief Medical Officer. The campaign resulted in the reduction of mortality rates from 300 during the yellow fever epidemic to 50 in 1904. By 1906 the authorities were able to claim that yellow fever was no longer epidemic in Rio. These findings had another practical implication in the area of international maritime commerce in that they made the use of quarantines and disinfection of merchandise irrelevant as tools for combating the spread of yellow fever. Thanks to the new knowledge generated by the scientific mission in Rio, the French public health authorities were able to design more efficient sanitation programs for Senegal and their possessions in the Caribbean.[16]

## The Application of the New Knowledge in Saint-Louis

On January 26, 1902, Governor General N. Ballay died after less than two years in Senegal. He was replaced by Ernest Roume, whose education, received at the *École Polytechnique*, made him the ideal person to apply the new knowledge in the framework of a program of *mise en valeur*, comprising public works such as railroads, ports, and health services. His previous work experience as Director in the Ministry of Colonies had helped him make connections within the financial circles in Paris. The execution of the sanitation program in Saint-Louis would take place under his governorship.

Roume had three legislative tools in his hands starting in 1903. The first, the ministerial decree of January 7, 1902—not promulgated in Senegal until 11 July 1903—made compulsory the declaration of 14 epidemic diseases, and mentioned in passing the responsibility of the medical personnel in reporting the diseases to the governor, the mayor or their representatives.[17] The targeted diseases included the following: typhoid fever, typhus, smallpox, scarlet fever, diphtheria, military sweating sickness, cholera and by-product, plague, yellow fever, diagnosed dysentery, puerperal infections, eye infection for newborn, measles, and leprosy.[18] The second law, aimed at protecting public health, was enacted on February 5, 1902. It defined the process of reporting the diseases as well as the sanitary measures to be taken against specific epidemic diseases, such as yellow fever, plague, and cholera. Doctors, heads of households, landlords, and *Canton* or village chiefs were require to report, within 12 hours, cases of epidemic diseases to the local administrators, mayors, secretary general of the government, and Chief Medical Officer. In cases of emergency, they were authorized to take the necessary sanitary measures, such as cleansing, isolation of the sick, and quarantine and disinfection before informing the authorities. The burial of the corpses was to take place within 24 hours.[19] Saint-Louis Mayor L. Descemet signed an ordinance with provisions for cleanliness, security and easy circulation on city streets, tranquility and the respect for public order.[20]

Under Roume's leadership the Governor General of French West Africa obtained fiscal autonomy and a centralized bureaucracy. Indeed, the decree of October 1, 1902, reorganizing the Governor General of the French West Africa, provided Roume with more control over the financial resources of the colonies.[21]

The third legislative tool in Roume's hand was the decree of April 14, 1904, which strengthened the provisions of the law of February 5, 1904. It addressed the issues of compulsory declaration of epidemic diseases and unsanitary housing. One provision of the decree created a Superior Committee of Public Hygiene and Sanitation, which was charged to advise the governor general concerning sanitary laws. Even if the decree did not resolve all the issues, it nevertheless created a general framework and gave the governor general and the governors of different colonies in French West Africa the full discretion to implement it and follow-up on the measures according the particular circumstances of each colony.[22]

The person directly responsible for Senegal was Governor Camille Guy, however. He had specific ideas about the implementation of the provisions of the decree of April 14, 1904. In a letter sent to Governor General Roume, dated May 15, 1904, Camille Guy recognized that, until then, the authorities never really sought the opinions of the members of the Hygiene Commission and the Municipal Commission, that their very attributions were ill defined and their roles limited, and that the appointment of new members was never on schedule. Guy went on to reaffirm his willingness to implement and follow-up on the new public health measures aimed at combating the 'infectious diseases and the insect that transmits them,' with the help of the representatives of the elected bodies and the notable residents. He committed himself to the organization of the campaign against the mosquitoes 'by applying the measures prescribed by the modern science,' which had proven efficient in Cuba, Brazil, and some foreign colonies against the same 'disaster.'

Guy's anti-yellow fever campaign would focus on the elimination of stagnant water everywhere: on the city streets, in the gardens, the courtyards, and houses, and on the cleanliness of the streets and docks, the draining away of rain water, the elimination of garbage and filth, in short, on the general sanitation. The campaign would be led by the new bureaucratic structures to be established, including the Municipal Hygiene Commissions and the Local Hygiene Committee. The agents in charge of implementing the sanitation program would be sworn in and authorized to penetrate inside private houses. Senegal would be divided into three sanitary districts, each monitored by a Special Commission.

Governor Guy warned Governor General Roume about the possibility of negative responses from some segments of the city residents to the implementation of the new sanitary measures. He classified them into three categories of people: 1) the partisans of civil liberties, who considered all the measures taken by the administration an attempt to reduce their freedom; 2) those who would interpret the new measures the administration's interference in their own affairs and habits; 3) and the 'skeptics,' 'the most dangerous group, who would laugh at and oppose a systematic resistance to the new measures and, thus, attract the sympathy of the naive and unhappy people around them.' But he would rely on civil servants, on people who believed in the necessity of public health measures, and the 'people of good faith'; and his approach would be through persuasion. He admitted that a simple decree 'would not modify the *indigènes*' old habits acquired over centuries, and lead them to abandon their prejudices, and their contempt for hygiene and hut cleanliness.' He concluded that the task ahead would require patience.[23]

As clearly laid out in this correspondence, Governor Guy's sanitation program was inspired by the new medical knowledge and the anti-yellow fever measures undertaken in Cuba and Brazil. It represented the end of the period of coexistence between the miasmatic and the germ theories of disease.

Guy also provides evidence for the persistence of the racial attitudes toward the urban poor. Although noncompliance came from all the segments of the population, including important sections of the colonists, Guy continues to associate the resistance of the urban poor to sanitary measures with their nature and culture. In the colonial discourse, it was natural for them to be dirty. As late as 1904, the 'culture of poverty' argument was still widely used in official circles.

In his response, dated May 18, 1904, Governor General Roume advised Guy not to create new bureaucratic structures, but to simply revamp the existing institutions (committees and commissions) established by the local decree of November 15, 1900 in accordance with the provisions of the decree of March 31, 1897; these institutions were in harmony with article 15 of the decree of April 14, 1904.[24]

When Guy submitted to the Local Committee of Hygiene and Public Health, presided over by Dr. Merveilleux, his projects of local ordinances to be issued in the framework of the decree of April 14, 1904, the suggested provision for the doctor or the sanitary agents to inspect private-owned buildings was rejected by 8 votes against and 3 in favor. The participants found the provision 'vexatious and not sufficiently justified' and invited Guy to rely on the existing decree related to the unsanitary and dangerous buildings.[25] Governor Camille Guy did not get discouraged by an apparent lack of enthusiasm exhibited by Roume and the members of the Local Committee of Hygiene and Public Health. In search of alternative solutions, Guy created the Municipal Hygiene Service,[26] and issued a decree related to the stagnant water in the interior of the cities. Residents were required to get rid of the discarded cans, vases and containers in and around their houses; and landlords were required to obtain the authorization of the mayor before undertaking any work of renovation in an old house or before building a new one. The decree also set the standards for safe housing.[27] Specific measures were taken concerning stagnant water in Sor and Bouetville.[28]

As *hivernage* approached, Governor Camille Guy promulgated a decree focusing on the measures to be taken in order to prevent or to stop the spreading of epidemic diseases. Besides the list of 14 diseases requiring a compulsory declaration, the decree had a provision for 8 diseases for which the declaration was optional: pulmonary tuberculosis, whooping cough, flu, pneumonia and broncho-pneumonia, *érysipèle*, mumps, ringworm, and conjunctivitis/eye infection.

One provision of the decree authorized the Health Service to disinfect the homes and furniture owned by individuals suffering from these diseases.[29] The rest of the decrees published in 1905 focused on various issues, such as smallpox vaccination and re-vaccination for children,[30] and housing conditions.[31] The Navy Ministry informed the Ministry of Colonies that, in case of re-emergence of yellow fever in Senegal, two ships would be ready in Toulon within 48 hours to repatriate in five days the non-essential personnel: the *Nive* could accommodate 822 people and the *Shamrock* had the capacity to transport 765 passengers. It would cost the colony 30,000 francs, without taking into account the expenses for charcoal and oil.[32] A few weeks before, the Ministry of Colonies had suggested that 1,800 to 2,000 Europeans could be repatriated, including 1,350 troops, 150 women, and 50 children to be evacuated during the first convoy.[33]

The legislation agenda testified to the state of preparedness of the administration in case of another epidemic crisis. The colonial authorities were determined to avoid the kind of panic that the 1900 yellow fever epidemic provoked. Governor General E. Chaudié's flight then could be associated with his perception of limited opportunities for escape. Now, the administration wanted to make sure that, by keeping completely open the routes for escape, an outbreak would not lead to panic.[34]

Between 1906 and 1912, the administration promulgated additional sanitary measures aimed at protecting the colony from imported diseases as well as locally induced diseases. The most important measures dealt with the sanitary maritime police and the protection of the French colonies from imported epidemic diseases,[35] the reduction to two years of the term limit for civil servants,[36] human trypanosomiasis[37]; the penalties for violating sanitary measures concerning the stagnant water in the cities, and other towns with European population.[38]

The success of these sanitary laws depended on the underlying scientific assumptions on which they were based and their enforcement. The evidence indicates that the number one preoccupation was the control of mosquitoes through the surveillance of the sanitary situation, disinfection, the destruction of the breeding sites for the mosquitoes, the use of mosquito nets, and the installation of the screens to the doors and windows of public buildings, such as garrisons, hospitals, and schools. Very often, the huts of the *indigènes* contaminated or suspected as well as the homes of the Europeans that were in bad shape and could not be disinfected were simply burned down. In order to limit the spread of infectious diseases, the separation between the Europeans and the *indigènes* was recommended where possible.[39] Public enemy number one was the *Stegomyia* mosquito.

Governor General William Ponty, who succeeded Roume in 1908, pursued the same sanitation program and the building of infrastructures. He provided the governors in French West Africa with specific instructions to combat the mosquito. He urged them to prevent the *Stegomyia* mosquito from reproducing; to destroy its larvae; 'to defend yourselves against it [mosquito] if it was not killed and to prevent it from becoming infectious.'[40]

The application of sanitary laws was characterized by moments of resistance as well as compliance. By the end of 1907 there were signs that the campaign against the mosquitoes was helped by the completion of the draining of the swamps on the island and part of Sor. The house-to-house visitation, undertaken in the framework of the mosquito control program led by the Hygiene Service, resulted in fewer cases of contravention, which could be interpreted as an acceptation by the majority of city residents, Europeans and Africans, of the Hygiene Service. The control measures intensified in November 1912 when cases of yellow fever were reported in Dakar and Kelle, a train station.[41]

There were also instances of resistance to the sanitary measures from ordinary people as well as from the bureaucrats. In early 1907, the *marabouts* from Guet-Ndar protested against what they considered the violation of their homes by the Senegalese sanitary agents. They were supported in their claims by the municipality, which underscored the persistent tension between the municipal authorities and health officials.[42] City residents also complained that the fees imposed by the tribunals on the violators of sanitary laws were lenient and cases of violation were not reported to the director of the Health Service.[43] There were also reports of corruption among the sanitary agents charged to enforce the sanitary laws. The agent Demba Seck is a good case in point. His area of operation was Ndar-Toute and Guet-Ndar and eyewitnesses testified that he regularly received gifts from residents who did not want their names to be reported as violators of sanitary measures. He was fired.[44] However, the ultimate success of the control measures depended on the elimination of the main breeding sites for mosquitoes constituted by the swamps in the northern part of the city and in Sor, the provision of pure water, and the construction of a sewage system. The following section will address the specific issues related to vector control.

## The Great Sanitation Program

The cost of draining the swamps in Sor and in the northern part of Saint-Louis, as calculated by the Sanitary Mission, was estimated at 1,650,000 francs.[45] The money for this project would come from the 5,450,000 francs destined to sanitation in Senegal out of 65,000,000 francs loan contracted in 1903 by the new Governor General Roume. The law of 5 July 1903 authorized the government of French West Africa to contract the loan.[46]

The sanitation program was granted to three builders or *entrepreneurs de travaux publics* (Touzet, representing Ed. Coignet; Sallenave; and Martinez), and part of the project was done under the system of '*Régie*,' for which the details were not provided. The most urgent among the projects was the draining of the swamps in the northern part of the island, the closest breeding ground for mosquitoes to the city. It required the transportation of sand from the dunes on the Langue de Barbarie to St.-Louis through barges and rail for 500,000 francs or 2.50 francs per cubic meter.[47] The adjudication for the construction in the northern part of the city and in Sor was granted to the *Entreprise* Touzet on March 22, 1905 for 347,266.5 francs, to the *Entreprise* Sallenave on July 29, 1906, and February 18, 1907 for 172,230.8 francs, and to the *Entreprise* Martinez on September 29, 1906 for 13,327.8 francs. The rest of the construction for part of the Pointe Nord and part of Sor was conducted in *Régie* for the cost of 285,174.7 francs. The total cost of the building of the infrastructure was estimated at 818,000 francs. The data shows that priority was given to the draining of swamps in the northern part of the island at the time when the mosquito control was the principal preoccupation of the administrations of Roume and Ponty (1902–1908; 1908–1915). The credits provided for the three projects were below the estimates, but all the credits made available were not even spent during the year. In 1909 the construction in Pointe Nord and part of Sor was completed, and the drained area was officially presented to the administration of the colony.[48] The maintenance expenses were estimated at 3,000 francs minimum per year.[49]

As the construction of the sanitation program on Pointe Nord was under way, health officials realized that the health of city residents depended also on the draining of the remaining swamps in Sor and the construction of the city's sewage system.[50] A certain number of initiatives can be understood in this context. In 1910 the Hygiene Service was authorized to extend its area of mosquito control to include Sor.[51] By then, it had created three anti-mosquito brigades. The first brigade operated since 1905 in the city-island under the supervision of an *agent de police*. The second and third brigades were operational only in 1911: the second brigade, led by a *gendarme*, controlled Guet-Ndar and Ndar-Toute, and the third brigade monitored Sor and was supervised by another *gendarme*. There was no brigade for the slum of Gokhumbaye. Besides the sanitary agents, prisoners were also participating in the cleaning of the stagnant water in the city.[52]

A project of the construction of the sewage system within fifteen months for the cost of 210,000 francs was also under examination in 1910.[53] Until then, city residents continued to throw waste matter, including human waste, into the river, thus increasing the risk of bacterial contamination. In the meantime, the *Entreprise* Sallenave obtained the authorization to extract 8,000 m³ of sand on the Langue de Barbarie for the embankment of the docks in the north and south of the city.[54] In 1911 the administration received an offer from the *Compagnie Française de l'Afrique Occidentale* to drain, at its own expenses, the swamps on its concession located southwest of the domain occupied by the railroad company in Sor.[55] The same year, the *Entreprise* Touzet was involved in the building of the docks in the northern part of the city.[56] When, in 1912, Governor General W. Ponty contracted 65,000,000 francs in loan for various public works projects, he provided 1,946,391.86 francs for sewage disposal and 1,628,602.35 francs for vector control.[57] Saint-Louis received 88,392.86 francs for its sewage system and 818,602.35 francs for the draining of swamps.[58] But in a letter to the governor general, dated May 28,

1912, Governor H. Cor argued that, according to the experts, the topography of the city did not allow an easy building of the sewage system, and that 261,000 francs would be used to drain the swamps in Sor and to provide more land to city residents. He suggested that the city would be better off with a modest system of wharf latrines built on the small arm of the Senegal River.[59]

The following year (1913) people connected with the municipal authorities expressed their dissatisfaction with the ways in which the administration was enforcing the sanitary laws. Indeed, on August 4, 1913, Dr. L. Huot informed the governor that the Local Hygiene Committee members, a group presided over by Mayor Justin Devès and dealing with urban hygiene, wished to see the draining of the swamps in Sor completed as soon as possible, and the enforcement of the decree of January 13, 1905, done with moderation until the swamps in Sor were drained. The request reflected the view that the mosquitoes in Sor marshes transmitted malaria, and not yellow fever, and that the dangerous mosquitoes responsible for the transmission of yellow fever, the *Stegomyia*, were highly domestic and lived indoors or near human habitation.[60] In other words, given the fact that the breeding grounds for mosquitoes on the city-island were eliminated, the Hygiene Service did not need to lead an aggressive anti-mosquito campaign in the city, but in the slums instead. From the perspective of the Committee members, malaria seemed to be less dangerous than yellow fever.

The pressure exercised by the Local Hygiene Committee members on the governor could also be justified by the presence in 1913 in the city of around 100 displaced fishermen and other poor residents from Guet-Ndar and Ndar-Toute, who were made homeless by the floods that destroyed their neighborhood in April 1912. The Committee members were aware of the reports from the Public Works Service indicating that about 250,000 francs were still needed to complete the work in Sor before displaced people as well as the surplus population (*population flottante*) could be accommodated.[61] By 1914 the displaced people were eventually granted some land concessions in Sor. But the First World War interrupted the sanitation program as well as other public works projects.

## Conclusion

The high mortality rates caused by yellow fever in Saint-Louis and other towns in Senegal between 1878 and 1881 gave a headache to the public health authorities and forced them in 1884 to adopt drastic sanitary measures aimed at containing the spread of yellow fever, the two most controversial being the designation of a vast coastal region of West and Central Africa as a 'diseased land,' and the imposition of an automatic annual quarantine of ships arriving from the targeted region during the rainy season between June and December. The model based on experience with miasmatic etiologies worked for 13 years before being modified in 1897 under the pressure from merchants operating on the West African Coast and the Navy officials engaged in colonial wars. According to its proponents, the model's predictions remained correct as long as the bureaucrats in charge of implementing sanitary measures fulfilled their duties with rigor.

The re-emergence of yellow fever in 1900 and the panic it created led some within the civil society to question the certainties of the miasmatic etiology at the time when Pasteurian microbial etiology provided an alternative explanatory model. But physicians and some of their followers remained skeptical about the new epistemology because it had not yet resolved the crucial questions of the identification of the infectious agent for yellow fever and its mode of transmission. Instead, they continued to rely on the miasmatic paradigm that had protected the city and its residents for 15 years. Nevertheless, the new discoveries to date did not support the quarantine system and its supporting mechanism of disinfection, which formed the backbone of the old paradigm. The 1900 yellow fever provided a platform where both the miasmatic and microbial etiologies clashed in the context of the debate over the best public health policy to adopt. In its aftermath, the French government sent in February 1901 a mission of inquiry to Senegal to find the causes of ill health and to make recommendations. By then, crucial experimentations taking place in Cuba and Brazil encouraged the French government in November 1901 to send a scientific mission to Brazil. The new findings transformed the knowledge about yellow fever that called for the adoption of a new public health policy based on a sanitation program and vector control. The microbial etiology would serve public health authorities well during the next major epidemic crisis when a bubonic plague epidemic struck the city in 1917. How can one account for the continuity and change in both the medical discourse and practices? The story of the bubonic plague will be told in chapter 6.

## Notes

1. *L'Écho de Paris*, no. 5943, Wednesday 5 Sept. 1900, 1–2.
2. Dr. Charles Carpot, *La fièvre jaune. Epidémie de l'année 1900 à Saint-Louis. Observations— Traitement* (Saint-Louis: Imprimerie O. Lesgourgues, 1912), 12, 15–16.
3. ANS/H49/AOF/9, Dr. Grall to Ministry of Colonies, no. 30 of March 6, 1901.
4. ANS/H49/AOF/9, Dr. Grall to Ministry of Colonies, no. 30 of March 6, 1901.
5. ANS/H48/1-2-3, Mission Sanitaire du Sénégal. Rapports 1901: Rapport d'ensemble; Hygiène militaire; Rapport partiel: hygiène urbaine; Rapport technique, Résumé et Conclusions. See also H 49 piece 41, Mission Sanitaire: Rapport sommaire, by Dr. Grall.
6. ANS/H48/1, Mission Sanitaire du Sénégal. Hygiène Militaire, by Dr. Grall.

7. Ibid., 24–25.
8. See François Delaporte, *The History of Yellow Fever. An Essay on the Birth of Tropical Medicine* (Cambridge, MA: M.I.T. Press, 1991), x.
9. ANS/H48, Mission Sanitaire du Sénégal, Rapport Partiel: Hygiène Urbaine, by Dr. Marchoux, 37.
10. ANS/H48/3, Mission Sanitaire du Sénégal, Rapport Technique, May 28, 1901, by Jacquerez, 53 p.
11. *Journal Officiel du Sénégal*, May 25, 1901, decree of May, 21, 225. The decree, issued by the Acting Governor General Lanrezac, defined the plan and conditions under which troops would be dispatched in camps, such as N'Diago, Pointe-aux-Chameaux, M'Pal and Louga for St.-Louis and Madeleines I and Ouakam for Dakar.
12. ANS/H49/AOF/66, minutes of a special Commission meeting, June 1, 1901.
13. ANS/H49/AOF/41, Mission Sanitaire: Rapport Sommaire, s.d., by Grall.
14. Ibid.
15. ANS/H49/AOF/67, Theodore Carpot, president of the *Conseil Général* and Deputy-Mayor, to Governor General, N. Ballay, September 25, 1901.
16. D. Tran, C. Chastel, and A. Cenac, "Paul Simond et la mission Marchoux au Brésil," *Bulletin de la Société de Pathologie Exotique*, 1999. See also Ilana Lowy, "Yellow Fever in Rio de Janeiro and the Pasteur Institut Mission (1901–1905): The Transfer of Science to the Periphery," *Medical History* 34 (1990), 144–163. E. Marchoux, Pl Simond, A. Salimbeni, "Etudes sur la fièvre jaune," premier mémoire, *Annales de l'Institut Pasteur* 17 (Nov. 1903, 665–728); Idem., 2me mémoire, *Annales de l'Institut Pasteur* 25 (Jan. 1906), 16–40.
17. *B.A.*, 1903, 361–62.
18. *B.O.*, 1903, 401.
19. *B.A.*, 1903, 401–4.
20. *B.A.*, 1903, 406–13.
21. See Margaret O. McLane, "Economic expansionism and the shape of empire; French enterprise in West Africa, 1850–1914," Ph.D. dissertation, U. of Wisconsin-Madison, 1994, 694; Alice Louis Conklin, "A mission to civilize: ideology and imperialism in French West Africa, 1895–1930," Ph.D. dissertation, Princeton University, 1989, 53–54.
22. *B.A.*, 1903, 272–78.
23. ANS/H19/AOF/8, Governor Camille Guy to Governor General E. Roume, no. 292 of May 15, 1904.
24. ANS/H?/AOF/14, governor general to governor, 18 May 1904.
25. ANS/H19/AOF/10, minutes of the *Comité Local d'Hygiène et de Salubrité Publique* meeting, May 30, 1904
26. *B.A.*, 1904, decree of January 5, 1905, 272–78.
27. ANS/H23/Senegal, decree related to the stagnant water, January 12, 1905.
28. *B.O.*, 1905, 281.
29. *B.A.*, 1905, 562–71.
30. *B.A.*, 1905, decree of 20 June, 741–44.
31. *B.A.*, 1905, decree of 4 July, 649–766.
32. ANS/H13/AOF/5, Ministry of Navy to Ministry of Colonies, 19 June 1905.
33. ANS/H52/AOF/2, governor general to Ministry of Colonies, no. 605 of May 16, 1905.

34. For the details on panic, see Neil J. Smelser, *Theory of Collective Behavior* (New York: Free Press of Glencoe, 1963), 131–69.
35. *Journal Officiel du Sénégal*, 1910, 70–79.
36. *Journal Officiel*, 1910, decree of 15 February 1910, 128.
37. *Journal Officiel*, 1911, August 29, 1911, 742–43.
38. *Journal Officiel*, 1912, 36–46.
39. ANS/H60, Governor General, Clozel, to Governor, no. 611 of April 3, 1912.
40. ANS/H54, no piece number, governor general to governors, November 30, 1911.
41. ANS/H23/Senegal, decree of November 17, 1912.
42. ANS/H32/Senegal, Annual Report 1907. The tension between the municipal authorities and the health officials had to do with the fact that the first believed that health officials were severe in the implementation of sanitary laws and that such severity was not justified, while the latter attributed the poor state of city sanitation to the inability of the municipal authorities to enforce city ordinances.
43. ANS/H13/AOF/196, governor to governor general, May 26, 1913.
44. ANS/H8/Senegal, Head of Health Service to governor, no. 42 of February 15, 1913.
45. ANS P 165/10, minutes of the *Comité des Travaux Publics* meeting, November 9, 1903. The cost included the elevator engine (90,000 francs), the draining of the swamps in Sor (1,121,000 francs), the draining of the swamps in the northern part of the city (297,634.43 francs), and miscellaneous expenses (141,909.87 francs).
46. Margaret O. McLane, "Economic expansionism and the shape of empire," op. cit., 194.
47. ANS P 165/8, director of Public Works to Governor, 22 July 22, 1904.
48. ANS/P165/AOF/36, Travaux Publics. Assainissement des Marais de Saint-Louis. Remise à la Colonie des Travaux Faits sur Fonds d'Emprunt. Situation des Dépenses Faites, by the Director of the Public Works, December 24, 909.
49. ANS/?/39, Rapport du Chef d'Arrondissement, Michas, no. 1206 of December 15, 1909.
50. ANS/2G6/24, Annual Report, 1906, 145–47.
51. ANS/G10/26, Annual Report 1910, 60.
52. ANS/H 60/Senegal, Bouet to Inspecteur des Services Sanitaire Civils, no. 18 of May 3, 1912.
53. ANS P 166/3, Travaux Publics. Emprunt de 65 million. Assainissement de la ville de Saint-Louis.
54. *J.O.*, 1910, 586, decree of 25 July authorizing Sallenave to extract sand on the Langue de Barbarie.
55. *J.O.*, 1911, 491, decree making public the adjudication of the draining of a piece of land in Sor.
56. *J.O.*, 1911, 407, decree of 2 June authorizing Touzet to occupy a piece of land for the duration of the work.
57. ANS/P/166/31, June 7, 1912.
58. ANS P 166/31, Finance director for French West Africa to Public Works inspector, no. 930 of June 7, 1912.
59. ANS P 166/32, governor of Senegal to Governor general, no. T 674 of May 28, 1912.
60. ANS/H8, Dr. L. Huot to governor, August 4, 1913.
61. ANS P 165/84, governor to governor general, no. ga 278 of March 12, 1912.

CHAPTER SIX

# Plague and Violence in Saint-Louis-du-Senegal, 1917–1920

The outbreak of the bubonic plague in Saint-Louis and the unprecedented and intrusive sanitary and medical measures adopted by the colonial officials to stop its spread challenge the received wisdom that the germ theory brought enlightenment to medical practice at "home" and in the colonies. The French public health officials specifically identified plague with the body of the colonized and focused their efforts on bodily apprehension and control, while the resistance of the urban poor to anti-plague measures equally centered on bodily evasion and concealment. Why the 1917 epidemic of bubonic plague provoked violence in Saint-Louis is the focus of this chapter.

Bubonic plague is a disease caused by the bacterium *Yersinia pestis* that is transmitted to humans through the bites of infected fleas, such as *Pulex irritans*, that have fed on infected animals, such as "squirrels, marmots, prairie dogs, mice, house cats, and, classically, the common domestic rat, *Rattus rattus*."[1] Human plague occurs after infected fleas run out of rodent hosts and start feeding on humans. Within six days of infection 60% of bubonic plague victims develop a bubo, or a grossly swollen lymph node sometimes reaching the size of an egg or an orange in the groin, armpit or neck, accompanied by a high fever, headaches, muscle pain, and mental disorientation; between 40 and 65% of infected humans succumb to the disease within a week after the appearance of the bubo. In some instances, varying between 5 and 15% of the cases, when the victim's lungs are infected, the bubonic plague can be transformed into a "primary" pneumonic plague, spread directly from person to person, and kill 100% of those infected. In other cases, the infectious agent spreads too rapidly for the victim to develop a bubo, leading to a "septicemic" or blood-borne plague. The majority of plague victims were children, the elderly, and the economically disadvantaged.[2] Myron Echenberg pointed out that people who live in "conditions of poverty, overcrowding, and malnutrition are at greater risk than others."[3] The outbreak of plague seems to be linked to the high-moisture weather that contributes to the multiplication of rodents.

Historians have distinguished three plague pandemics. The first pandemic ravaged the Mediterranean region between the fifth and eighth centuries. The second pandemic visited the entire European continent where it created panic and terror between the mid-fourteenth to the end of the eighteenth centuries, killing around 25 million people between 1348 and 1352 and leading to a witch hunt against Jews. The third and last pandemic started in China in 1894 and spread all over the world, producing 13 million victims in India alone between 1898 and 1948.[4] The 1896 plague epidemic in Bombay reached the African continent through international travel and trade: Tamatave (1898), Alexandria (1899), Durban (1903), Cape Town (1900), Abidjan (1899), Dakar (1914), and Saint-Louis (1917). Plague continued to spread in Senegal and cause mortality until 1945.

The bacterium was discovered in 1894 in Hong Kong by physician A. Yersin, a Swiss naturalized French who contributed to setting up the Pasteur Institute and to teaching microbiology before working as physician resident aboard the "*Messageries maritimes.*" Yersin is also credited for uncovering the role of rats in the transmission of bubonic plague, and, in 1896, the anti-plague serum, while another Pasteurian, Paul-Louis Simond, made in 1898 in Karachi a fundamental discovery in the epidemiology of plague, that is, the role of fleas in the transmission of the bacillus that was the missing link in Yersin's work.[5] The operations of deratisation and disinfection of ships as well as the building of entrepôts and sewers inaccessible to the rodents contributed to reducing the spread of plague throughout the world. The treatment consists of sulfamides and antibiotics.

Myron Echenberg has helped understand additional factors that contributed to the spread of plague in Senegal. Some occupations exposed individuals to the disease, including the dock workers, grain handlers, and bakers who entered into contact with rodents that fed on grain supplies. Women who were more sedentary than men were also at more risk than men. Also the introduction of peanuts as the main commercial crop in Senegal provided a contributing factor for the spread of plague. The peanut basin became also a plague zone.[6]

## Outbreak in Saint-Louis and Political Response, September–December 1917

In early September 1917, there were rumors among the French residents about cases of plague observed in Saint-Louis and the apparent spread of the disease in the peri-urban villages. The medical authorities were disappointed that the epidemic preparedness undertaken in 1914 to protect the city against the Dakar plague outbreak was ineffective. Then, they had chosen the Gardettes Building to house the eventual patients, opened a lazaretto on Baba-Gueye Island for plague suspects, regulated the movement of floating population, kept a watchful eye on the overcrowded peri-urban villages and unsanitary dwellings, encouraged city residents to destroy rats by offering 0.25 cents per captured rat, and offered the Haffkine vaccine on a voluntary basis.[7] The circulation of rumors of an epidemic in the city three years later provoked fear and panic among city residents even if the authorities had not yet officially announced the outbreak of plague.

It was not until 7 December that the governor confirmed the existence of clinical cases of plague in the city. The focus of infection was located in Ndar-Toute and around Sidy Tall Mosque. The authorities took precautionary measures to stop the spread of the disease in other parts of the city and the rest of Senegal. The decision was a calculated move on the part of the colonial officials, who had at their disposal panoply of urgent sanitary and medical measures within the existing disease surveillance system, as defined by the ordinance of 16 July 1903, in order to combat the spread of the disease. As in previous epidemics, the authorities were hesitant to create panic that could disturb commerce and paralyze the administration. But as the disease spread, the colonial authorities decided to intervene.

According the provisions of the ordinance of 7 December, city residents were required to notify the colonial authorities about new plague cases. The ordinance gave the medical authorities powers to isolate the sick for medical care, and put people who were in contact with the patient and were considered plague suspects under medical surveillance in the lazaretto at Pointe-aux-Chameaux for a period of 10 days; to destroy or disinfect suspected dwellings, furniture, and other suspect objects; to inspect and issue an unclean bill of health to any ship leaving Saint-Louis; and to prohibit the importation of any product susceptible of transmitting the disease, such as skins, used objects, bedclothes and rags that carried infectious agents. *Indigènes* as well as foreign minorities (Syrians and Moroccans) present in Saint-Louis were required to hold a health passport or medical certificate issued by the sanitary authorities showing proof of a completed quarantine or vaccination. Travelers without this medical document would be denied access to the train station.[8] Vaccination was made available to city residents.[9] The declaration of an epidemic meant that the colonial state would intervene into more and more aspects of people's lives. Physicians were authorized to make daily inspection visits in the houses located within the infected perimeter in search of people presenting the symptoms of the disease. People would be told what to do and where to go or not to go.

But three days later, on 10 December, the disease acquired a specific social and spatial character, as the residents of Ndar-Toute and Guet-Ndar became the targets of the anti-plague measures. Indeed, the governor imposed a sanitary *cordon* on the two peri-urban villages, and restricted the contacts between them as well as between them and the rest of the city; targeted residents were prohibited to change residence without an official authorization.[10] In order to prevent the clinical cases and people who had been in contact with them from spreading the disease, residents were required to notify the municipal medical authorities of cases with suspected plague symptoms as well as suspected plague-related deaths.[11] A three-member commission was put in place to evaluate the cost of infected property to be destroyed.[12]

## Popular Protest

Resistance took various forms ranging from violent protests to peaceful demonstrations and passive actions. Some leaders of violent protests were often arrested and prosecuted. The most popular form of peaceful protest was the refusal by the women of Guet-Ndar to sell fish at the market.[13] The first protests started in December when physicians Lailheurgue and Le Gallen visited a plague suspect, daughter of Amadou Moctar, in Ndar-Toute and were told that she went to the city-island. At the same time, women of Ndar-Toute took to the street to express their anger against the specific restrictions imposed upon them, and verbally abused the two doctors as they were leaving the premises, prompting the resignation of Dr. Lailheurgue.[14] The next day, health officials extended the vaccination that started on 7 December to the rest of the city and replaced the sanitary passports with vaccination cards. The same day, François Carpot, the *métis* deputy and lawyer, conveyed the grievances of the people of Ndar-Toute to interim *Secrétaire Général* Muller concerning the harshness of the anti-plague measures and the unfair compensation for destroyed property.[15]

The administration officials believed that, although the sanitary and medical measures adopted were not sufficiently "energetic," their strict application would have contained the epidemic. But most provisions of the early anti-plague measures never received the beginning of execution, especially in Guet-Ndar that was considered "the main focus of infection" because of a number of factors, chief among them the resistance of the people of Guet-Ndar and the loopholes in the legislation that made the prosecution of protestors difficult.[16] In the meantime, the ravages of the epidemic continued to spread and, by the end of 1917, plague mortality was estimated at 58 deaths.

Resistance resumed in mid-January when the first two deaths in Guet-Ndar were followed by the evacuation of only two individuals to the lazaretto, the other plague suspects refusing to be isolated. In a telegram dated 16 January 1918 to the governor in an inspection tour at Bakel in Upper Senegal, interim *Secrétaire Général* Muller reported that the entire population of Guet-Ndar had exhibited a "bad temper," threatened the Europeans, and resisted all attempts to remove the corpses of plague victims and evacuate the suspects, thus forcing the Hygiene Service to suspend its activities in that peri-urban village. He also mentioned another serious incident that took place in the northern *quartier* of Saint-Louis where two European gendarmes—Bouville and Perennsez—and several European and African health workers were beaten up as they tried to remove the body of Bouna, an interpreter of the tribunal who had passed away the night before. He described the situation as a "worrisome effervescence," requiring the use of armed force, and urged the governor to return to Saint-Louis where plague cases were in a "marked recrudescence," as the first European plague case was reported and sent to the lazaretto.[17] The colonial authorities' anxiety was justified in the light of the publication of the fatality case for the first two weeks of January which indicated a total of 22 new plague cases and 14 deaths. Ndar-Toute had the highest number of cases (15 cases with 11 deaths), followed by Saint-Louis (4 cases) and Guet-Ndar (3 cases with 3 deaths).[18] But the people of Guet-Ndar had their own perspective on the causes of the tension. In a letter to the governor dated January 18, 1918, they made it clear that they did not reject the medical advice: "What we disdain," they argued, "is to see our dead taken away by the hygiene service when we can ourselves bury them as well...according to the requirements of our religion."[19]

It should be emphasized that the colonial authorities came to construct resistance as a "Wolof resistance," while the Bambaras (Bamana) were described as "docile," accepting to be evacuated to the lazaretto "without any resistance." Such a perception led Dr. Gallen of Hygiene Service to wonder whether the colonial authorities could redefine the "native problem" in Senegal in terms of "races" instead of religion, as both Wolof and Bamana were Muslims and yet responded differently to the European demands, and to question the relevance of the usual colonial categories (Europeans, *assimilés*, and *indigènes*).[20] Moreover, the resistance of Guet-Ndar residents was seen as a contagious disease capable of spreading and infecting (healthy) people in other parts of the city.[21] But the evidence suggests that violent incidents were first signaled in Ndar-Toute and Saint-Louis, at the time when there was still calm in Guet-Ndar. In addition, there were instances when the Bamana resisted the evacuation to the lazaretto.[22]

A serious incident took place at the occasion of the death, on 22 January 1918, of Coumba Diaw, a woman of Guet-Ndar who had shown symptoms of bubonic plague. Once notified of the death, Dr. Damian, Head of the Health Service, invited Inspector Mailleraud and two other municipal doctors to visit the Lodo *quartier* in Guet-Ndar in order to provide *marabout* Aldia Gueye with specific instructions for avoiding infection during the burial ritual, especially the ritual washing of the corpse, and to isolate in lazaretto the suspected cases, that is, those who had been in contact with the victim. This first encounter between the medical authorities and the residents of Guet-Ndar revealed a great deal of the mutual suspicion that existed between the mostly fishermen and the colonial officials as a result of past experience. Indeed, the medical team noticed that the 7,000 residents of Guet-Ndar were expecting their visit and were prepared to resist the sanitary and medical measures taken by the state officials. Nobody wanted to provide the medical team with the vital information they needed, especially the identification of suspected cases to be isolated. Plague patients fled their huts. The "mob" started to gather around the medical team and Faly Sene, a local notable, shouted that "*tous refusent de se laisser évacuer et (que) personne ne bougera*" (everyone refuses to be evacuated and (that) nobody will budge). As the tension rose, health officers left the *quartier*. The "mob attack" on them ended only when they reached Servatius Street at the entrance to the *quartier*.[23] So the medical team left Guet-Ndar without seeing Coumba Diaw's body or granting permission to bury her, or applying other precautionary measures such as the isolation of the suspected cases, the burning or disinfection of her home with sulfuric acid, and deratisation.

## Alternative Strategies: Biopolitical Actors vs. Sanitarians

This first encounter between the medical authorities and the urban poor only reinforced state assumptions about its Civilizing Mission, which were translated into plague policy, and about the "backwardness" of the urban poor. But there was a division within the administration between sanitarians (medical authorities) and biopolitical actors (public officials) concerning the attitude to adopt vis-à-vis the popular protest to sanitary and medical measures in the early phase of the plague epidemic. The medical authorities, who had observed signs of popular violence in Guet-Ndar, argued in favor of the strict application of the provisions of the ordinance no. 2093 of December 12, 1912 related to the compulsory declaration of epidemic diseases, which were being implemented and which contained provisions that violated the privacy of people's homes. Physicians wanted the administration to suppress the protest and establish the "rule of law." Public officials (governor, governor general, administration officials), in contrast, seemed to prefer accommodation to conflict. Governor Lévecque contended that the reaction of the people of Guet-Ndar to sanitary and medical measures could be understood as a reflection of their "ignorance" and in the context of the Islamic religion. He suggested an approach that would convince the moderate *marabouts*, such as Marseck from Ndar *quartier* in Guet-Ndar, Aldia Gueye from Lodo *quartier* in Guet-Ndar, Amadou Sarr, and Diaye Sarr Elumane from the mosque, as well as the *cadis* (judges at Muslim tribunals), who were the natural leaders of the Guet-Ndarians, about the compatibility of the sanitary measures with the prescriptions of the Koran.[24] Some of the contentious issues concerned the widespread ritual burial, especially the washing of the corpses, the burial ceremony and the mourning practices, which attracted crowds and contributed to the spread of infectious agents. It was clear that Lévecque was not prepared to follow the "strict application of the indispensable sanitary measures" as health officials had hoped. Initially, the views of colonial officials seemed to prevail.

## Carrera's Intervention

Governor Lévecque solicited the intervention of members of the leading families in Saint-Louis who had immense prestige among the lower classes and spoke Wolof. He persuaded Carrera, Administrator of Colonies, to act as mediator between the colonial state officials and the mostly fishermen from Guet-Ndar. Carrera accompanied Dr. Damian, Chief Medical Officer, and Dr. Dupont and his staff at the municipal Hygiene Service, to Guet-Ndar on January 27, 1918 in order to convince the local notables to comply with the anti-plague measures related to the isolation of plague suspects, the disinfection or burning down of infected dwellings, the ritual washing of the corpses, and other burial rituals, and the request for burial permits. But the delegation from the city-island failed to convince the notables from Guet-Ndar to comply with the plague policy as defined by state officials. The local notables simply wanted "to be left alone with a disease that they accepted with the fatalism characteristic of the Muslim religion." At the end of January, 12 cases with 12 deaths were reported in Guet-Ndar. As of 9 February, 9 cases with 9 deaths were also reported. Doctors expressed the fear that, given the fatality rate of 100% observed until then, the swarming of the fleas in the next couple of months could transform the already overcrowded peri-urban village into a dangerous "focus of infection."[25] Mediation was needed.

## Blaise Diagne's Intervention

In the last attempt to avoid the use of force, colonial officials sought and obtained the intervention of the *Commissaire de la République*, Mr. Blaise Diagne, the first black to be elected *député* to French Parliament in 1914. On 21 February 1918, Diagne sent a telegram to the mayor of Saint-Louis urging him to get involved in the resolution of the medical crisis. He wrote the following:

> I appeal to your high conscience in order to attract the attention of our compatriots from Guet-Ndar on the imperious and urgent necessity for them to accept all the administrative measures that were taken or would be decided in order to wipe out the plague stop. Nobody would accept that because of lack of reason our compatriots would contribute to the spread of the epidemic disease which would ravage Saint-Louis as well as the entire colony and perhaps the entire FWA.
>
> At the time when military concentrations will take place following the recruitment of natives any hesitation in the preservation of public health will be tolerated no matter the cost.
>
> I would thus be very grateful if you would apply an energetic and last pressure on Guet-Ndar by gathering the local notables of the village along with the leading residents of Saint-Louis in order to make them understand and accept all the measures that the administration has adopted and will adopt.
>
> The deadly character of the disease requires these measures that, however painful they may be, remain compatible with the principles of the Islamic religion stop—If I, son of this land, exhort my compatriots to submit to the rigors of the situation it is because only there lies the preservation of all the population.[26]

The fact that the colonial officials bypassed the mayor in dealing with the Guet-Ndarians strongly indicates the existence of some form of tension between the municipal authorities and the colonial authorities. Indeed, during previous epidemics the administration officials blamed the successive mayors for the filthy state of the city and the peri-urban villages. An appeal to other members of the African elite must be understood in this context. Diagne's argument was not different from the "culture of poverty" explanations promoted by the administration, which viewed the Africans as ignorant and irrational. He did not take into account the urban poor's grievances. This is the reason why his intervention through the first elected black Mayor, Pierre Chimere (1916), had little chance of success.

In the meantime, the administration expanded the methods of inspection beyond simple police surveillance by sending troops to enforce the sanitary *cordon* around Saint-Louis, Ndar-Toute and Guet-Ndar in order to arrest the sanitary *cordon* violators by land and by river.[27] The authorities also imposed a quarantine of 10 days in the lazaretto on city residents planning to travel out of Saint-Louis while holding a medical passport. To obtain this document, the *indigènes*, Syrians and

Moroccans were required to first show proof of vaccination. In any case, they were not allowed to take the train between Saint-Louis and Louga to Dakar.[28] Fishermen were allowed to fish and sell fish under certain strict conditions in order to ensure the continued food supply to the city, all transactions taking place along the metallic fence. A pass from the Hygiene Service was required to cross the sanitary *cordon*. A burial procession in Guet-Ndar was reduced to a maximum of 10 people and had to be escorted by two soldiers. A military escort was also required for the maximum of three *indigènes* authorized to go to the city hall to request birth or death certificates for a relative, and for those invited to appear at the city hall, the police or *gendarmerie*.[29]

Governor General Angoulvant endorsed the comprehensive policy suggested by Governor Lévecque. He encouraged him to enforce the sanitary *cordon*, and find a peaceful solution to the crisis,[30] but without letting the rebellion against the police and hygiene service agents go unpunished. Given that all means of persuasion used by the administration, including Blaise Diagne's intervention, had failed, Angoulvant urged Lévecque to enforce the existing laws while avoiding a bloodbath.[31]

## From Accommodation to "Assault on the Body": The Colonial Hygiene Committee

By March the medical authorities made a last attempt to convince the colonial officials to abandon their accommodationist approach in favor of the use of force. The members of the Colonial Hygiene Committee, including five physicians, one veterinarian and five other members,[32] held an urgent meeting on 21 March 1918 in order to consider the strategies to deal with the health crisis. They unanimously agreed that "the free penetration of the Hygiene Service in Guet-Ndar to conduct their operations must be made possible by any means necessary, including the use of an armed force and no matter the consequences that would result from it," and that all the Guet-Ndar residents had to be classified, *en masse*, as plague suspects to be detained and evacuated in successive groups to isolation camps to be erected in Sor and on Langue de Barbarie south of the Pointe-aux-Chameaux lazaretto for a period of 10 days.[33] Mayor Pierre Chimère attended the meeting but did not question the harshness of the plague measures for reasons difficult to explain without additional evidence. Taken in a context dominated by the psychology of terror, this plague policy was not without risks for the administration. The committee's members did not even consider alternative solutions, such as face-to-face communication with the notables of Guet-Ndar, nor did they calculate the financial as well as the political cost its implementation would require. In their view the city's health had to be protected by any means necessary.[34]

In his report to the lieutenant-governor the following day, Dr. Contaut, who had replaced Dr. Damian as Chief Medical Officer, argued that the use of force was the best strategy to combat what he then perceived as an epidemiological shift from bubonic to pneumonic plague, which he believed was dangerously spreading from human to human from its "irreducible center" in Guet-Ndar. He provided the details of the new plague policy adopted by the committee members, which included the evacuation and detention, over a period of several months, of the entire population of Guet-Ndar in the isolation camps for a period of ten days for each group, the burning down of the infected huts that formed the majority of the dwellings, the disinfection of the few suspected houses built in durable materials, the relocation of everyone to a temporary segregation camp for further medical surveillance, and their final relocation preferably in several new peri-urban villages to be created. The confinement of the urban poor to one temporary location would provide the medical authorities with the opportunity to visit plague suspects, record deaths, and control burial rituals by putting antiseptic substances in the coffin. Guet-Ndar would be cleaned up with fire, "the only energetic and radical agent of sterilization."[35] The same day, the autopsy performed on the body of a battalion chief Petitjean revealed the septicemic plague as the cause of his death.[36]

Dr. Contaut's report reflected the widespread belief among French colonial officials and others that the urban poor had brought the situation upon themselves by putting themselves outside the common law; they were to blame for the spread of plague because of their "ignorance, lack of discipline, and undesirability." Dr. Contaut went so far as to label the urban poor, especially the fishermen of Wolof origin living in Guet-Ndar, as "a special race" with "a particular mentality." He prepared the ground for a justification of the special measures taken to combat the plague in the city. In his perspective the financial, social and political cost of the plague operation, including the complete suspension of individual freedom of citizens, would be minimal in the short-term comparing with the long-term cost that the suspicion of Senegal would create abroad if the epidemic was allowed to spread further from its dangerous hotbed. The general interest prevailed over individual liberties.[37]

Dr. Contaut did not solely focus on the plague policy; he also laid down the foundations of a new urban *politique indigène*, based on the fragmentation of the urban poor—who had acquired a kind of "class consciousness"—as well as the fluctuating population (Moors, Bamana, etc.) into several ethnic-based villages. The ultimate aim was to avoid in the future "the negative mental disposition that has characterized Guet-Ndar until now.[38] Dr. Contaut's reasoned argument, warning about the consequences for the colony of the implementation of an alternative strategy to combat the epidemic, was aimed at creating doubts in the mind of the governor of Senegal, Mr. Lévecque, in order to convince him to adopt the new proposed plague policy. What was the governor's response to Dr. Contaut's report?

## Governor Lévecque's Dilemma

Governor Lévecque rejected the radical plague measures—such as opening fire on the protestors, setting houses ablaze in Ndar-Toute and Guet-Ndar, and resettling everyone somewhere else—proposed by Dr. Contaut and his colleagues, who dominated the Colonial Hygiene Committee, even if they both shared the colonial state's perceptions of the urban poor. Indeed, in his long letter to the governor general of French West Africa dated 25 March 1918, Governor Lévecque described Guet-Ndar as "overcrowded," "with houses on top of each other," and he characterized the response of the Guet-Ndarians to sanitary and medical measures as a "firm resistance" compared to the "sporadic resistance" observed in other parts of the city. Lévecque nevertheless made a conscious effort to try to understand the reasons behind the Africans' response to anti-plague measures. He attributed that resistance not only to the "indigenous customs, superstitions, and religious beliefs and practices," but also to the encouragement that the protestors received from "unscrupulous individuals, for whom everything goes when it comes to conserving, regaining or winning the confidence of the voters one may need."[39] Clearly, a conspiracy theory was being developed within some administration circles to make sense of the reactions of the urban poor to sanitary measures.

Although Governor Lévecque and Dr. Contaut agreed on the causes of the crisis, they differed, however, concerning the solutions to the problem. From the governor's perspective, the suggested anti-plague measures were "radical," "excessive" and risky. He was aware of the fact that many among the European city residents disagreed with him, especially those "who are surprised that fire is not yet set on the four corners of Guet-Ndar, that no shots were fired at those people, that they were not thrown to the sea."[40] He had an answer for his critics: "I have not burned down Ndar-Toute where the epidemic disease tends to disappear; I have not yet burned down Guet-Ndar because I had to consider the consequences of that radical measure, and did not want to aggravate an already complicated situation."

Having set the record straight, Lévecque went on to present his alternative solution to the crisis. His views were more moderate and realistic, taking into account the imperatives of justice, reason, the available financial and human resources, especially the small size of the police available for the plague operation, and people's cultural and religious beliefs and practices. Moreover, Governor Lévecque was well aware of the fact that the majority of the urban poor living in Guet-Ndar were French citizens since 1848; he did not want their civil liberties to be violated in the name of public health and was prepared to oppose radical measures that would leave an embittered population. His efforts were paying off. Some leading resistors had been prosecuted for rebellion. Among them were Fergueye Gueye, Alioune M'Boye, and Makhary Samb. Gueye was charged with assault (*voies de fait*) but he was acquitted on 18 January for having acted without discernment. M'Boye of the southern part of Saint-Louis opposed the evacuation of people to the lazaretto; he was arrested and charged with rebellion and menace to the police inspectors during the exercise of their functions; he received a six-month jail sentence on 13 March. Samb was scheduled to appear for arraignment on 28 March. Thus, Governor Lévecque hoped that the combined effect of indictment and successful prosecution of some leading resistors as well as the pressure from the sanitary *cordon* around Guet-Ndar, in addition to the one around Saint-Louis, would create weariness and, eventually, would break down the protestors' morale to the point of surrendering to the police.[41]

The optimism of the colonial officials was justified in the light of the encouraging signs coming from the leaders of the urban poor. In early March, 33 moderate notables from Guet-Ndar communicated to the colonial officials their willingness to submit to the sanitary and medical measures and revealed to them 11 names of the leading protestors against the measures, including one former chief, Birahim Gaye.[42] The initiative reflected a split within the Guet-Ndarian community itself between moderates and radicals. There were reports of an increasing tension between the residents of the two neighborhoods, one group preventing the other from selling fish.[43] The leaders of the moderate faction also sought the intervention of François Carpot, former *député* (1902–1914) and a member of one of the most prestigious Creole families in Saint-Louis. On 7 March 1918, 28 "habitants de Guet-Ndar" sent him a letter requesting his assistance in convincing the colonial administration officials to lift the quarantine. They underlined the social (and geographical) character of the plague policy which discriminated against them simply because they were the most vulnerable segment of the urban population. They argued that such treatment was not appropriate in the light of the contribution they had made in the framework of the war effort in terms of troops.[44]

In addition, Moctar Bouna, Chief of Guet-Ndar, made contacts with the police chief to arrange a meeting with the municipal authorities in order to discuss the conditions of the application of the plague measures. However, the police denied him access to the mayor's office because he was only in favor of disinfecting the contaminated huts but opposed handing over the corpses to the medical team as well as the isolation of plague suspects.[45] Both sides missed the opportunity to resolve the crisis peacefully because Western prophylaxis was incompatible with the indigenous theories of contagious disease and practices. Groups involved on both sides were unable to effectively communicate cross-culturally.

## Mortality, Fear, Protest, and the "Radical Solution"

The publication, on April 3, 1918, of a special report on case fatality for March, confirmed the apprehensions of the medical authorities. There were 9 deaths reported on the city-island, 1 in Sor, 1 in Ndar-Toute, and 57 in Guet-Ndar in the previous two weeks (15–31 March). Mortality in Guet-Ndar was increasing compared with 48 deaths recorded at the beginning of the month (1–14 March), including 43 deaths due to pneumonic plague. The clean-up efforts had resulted in the capture of 852 common domestic rats, *Yersinia pestis* carriers, during the same time period.

The report showed that there was an improvement in the *quartiers* where regular disease surveillance, disinfection and burning down of infected or suspected dwellings and case notification took place and a high case fatality in Guet-Ndar. In his construction of the epidemic, Dr. Contaut attributed the high mortality in Guet-Ndar to "our inertia *obligée* vis-à-vis Guet-Ndar" and called for "a plan of action against this refractory village" that constituted "a dangerous thorn on Saint-Louis' side."[46]

The same day, 3 April, Governor General Angoulvant sent a telegram to Governor Lévecque in response to his letter dated 25 March. Angoulvant also rejected the radical strategy proposed by Dr. Contaut and the other members of the Colonial Hygiene Committee to combat the epidemic. Administrative and financial considerations weighted heavily in his decision. He urged that the sanitary *cordon* around Guet-Ndar remain in place and that the governor initiate contacts with local moderate notables in order to find if there were volunteers for 10-day quarantine at Pointe-aux-Chameaux, and give indemnity to the people who would lose their dwellings. He posited a linkage between compliance with vaccination and revaccination and the lifting of the sanitary *cordon*.[47] The task ahead was difficult given that at the same time there were reports that 12 Guet-Ndar residents, who had to appear before the instruction judge, categorically refused to first submit to medical examination at the lazaretto.[48]

The next report, made public on 17 April and covering the previous two weeks (1–15 April), gave 3 deaths in the Northern part and 3 deaths in the Southern part of the city-island, 1 death in Sor, 4 deaths in Ndar-Toute, and 43 deaths in Guet-Ndar. The report indicated a general improvement in mortality compared with the previous report. The improvement was due to two factors: 1) the predominance of the bubonic and septicemic forms of plague and the decline in cases of pneumonic plague, the most contagious form of plague, 2) and the migration of infected rats from Ndar-Toute to Guet-Ndar and from Ndar-Toute to the city-island. The medical authorities were able to establish Ndar-Toute as the point of departure of the epidemic and to link the movement of the infected rats to the clinical cases and

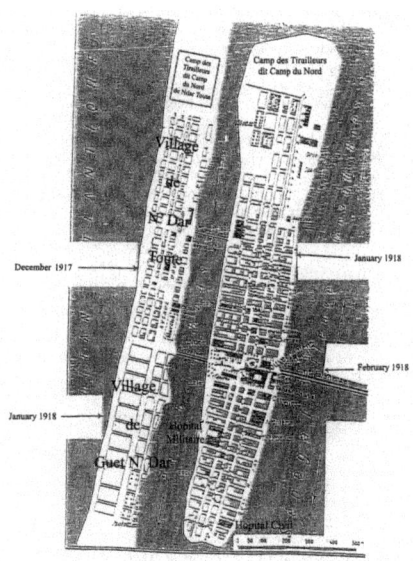

*Mage St. (January, February) (from N to S)*

↓

Ndar-Toute ⟶ Geole Bridge ⟶ City Place du Government (February)

↓

*Bisson St. (end of March)*
Ndar-Toute ⟶ Guet-Ndar (December, January)
Ndar-Toute ⟶ Sor (Dec.): visit by two people incubating the disease;
Guet-Ndar ⟶ Sor (Febr.): visit by infected people.

Map 4. *The Evolution of the Bubonic Plague Epidemic in Saint-Louis (adaptation from* La Geographie, *nos.4–5 (Nov.–Dec.). 1925. p. 424)*

deaths reported. From its basis in Ndar-Toute the disease spread with the movement of rats as shown above.[49]

The medical authorities could reconstruct the evolution of the plague epidemic with confidence. They explained the low mortality rate in Sor in relation to its particular ecology characterized by the abundance of sand that was an obstacle to the multiplication of rats.

In the meantime the medical authorities continued with the operation of deratisation on the city-island and in Ndar-Toute. To expand the sanitary law enforcement powers, the governor general requested the abrogation of article 31 of the local ordinance dated 12 December 1912 and its replacement with a new ordinance that would 1) increase the size of the Hygiene Commission of Saint-Louis to include the personnel already involved in the execution of the sanitary measures, such as physicians, police officers, gendarmery brigade chief, etc., 2) and toughen the penalties. From then on, refusal to comply with sanitary legislation would result in 15 days in jail and the payment of a penalty of 1,000 francs.⁵⁰ Angoulvant also moved to resolve the issue of conflict of interest faced by the mayors of Dakar and Saint-Louis, especially during the epidemic crises, by suggesting the promulgation of a new decree, issued on 6 May 1918, that delegated all prerogatives related to hygiene and public sanitation to another especially appointed civil servant.⁵¹

By May the crisis in Guet-Ndar had deepened. The supporters of the conspiracy theory, who saw a linkage between the popular protest in Guet-Ndar and the covert actions of some "political personalities," including "those who previously occupied high elective offices," hoped that Governor General Angoulvant would take radical measures to deal with the "serious incidents" reported in various quarters of the city as well as the continued defiance campaign against the plague policy led by 8,000 Guet-Ndar residents. But the governor general continued to support the strategy of accommodation adopted by Governor Lévecque. In a letter to the president of the *Conseil Général* dated 4 May 1918, Angoulvant argued that burning Guet-Ndar would only be a partial solution, for the colonial state would have to rebuild it as the majority of its residents were French citizens. He broached aside any attempt to establish similarities between Guet-Ndar and Dakar in 1914, where a new quarter (Medina) was erected for displaced people. He would only accept a temporary sanitary segregation but not the erection of several new permanent quarters to accommodate the residents of Guet-Ndar. He was in favor of a balance between civil liberties and health concerns, and of a new legal framework to deal effectively with the protest.⁵² There were also financial considerations as the expenses for plague operation doubled, passing from 25,531 francs in 1917 to 50,000 francs in 1918.⁵³

The publication, on 7 May 1918, of another special report on plague mortality in Saint-Louis for the second half of April (16–30 April) revealed 73 cases and 62 deaths, including 38 deaths in Guet-Ndar, 1 death in Ndar-Toute, 3 in Saint-Louis North, 4 in South, 3 in Sor, 1 in Gandiole, and 13 in the lazaretto. The actual number of deaths in Guet-Ndar was higher than the official statistics indicated, if one takes into account the clandestine burials taking place inside the dwellings. The report also established that the vaccination and re-vaccination of the population, the isolation of clinical cases and suspects, the disinfection of homes, and the payment of incentives for the destruction of rats had intensified. But the residents of Guet-Ndar did not show signs of compliance with the sanitary and medical measures,⁵⁴ even after the issuance, on 25 June 1918, of an ordinance making the anti-plague vaccination and re-vaccination compulsory in Saint-Louis, Dakar and the towns located along the railway between the two cities.⁵⁵ Given the difficulties to import vaccine from France, the administration encouraged Dr. Leger to produce vaccine in the laboratory in Dakar; the production of vaccine went from 10,000 cmc to

150,000 cmc per month.⁵⁶ Vaccination consisted of a double inoculation at five days of interval; and the re-vaccination took place five months later. A certificate was issued as a proof that the individual was vaccinated.⁵⁷ But more state intervention in people's lives led to more protests.

Protest continued through July into early August, while the leaders of Guet-Ndar multiplied contacts with the colonial authorities in Saint-Louis to reach a compromise but with little success. On 12 July, Lévecque submitted a plan for a progressive evacuation and disinfection of Guet-Ndar, and the *Général Commandant Supérieur* his military strategy.⁵⁸ On 1 August 1918, Governor General Angoulvant tried one more time to persuade Blaise Diagne to travel to Saint-Louis to bring the people of Guet-Ndar to their senses. The real problem, as the governor general understood it, had nothing to do with the compensation for destroyed property, but their declared intention to disobey the laws and not to cooperate with the colonial authorities in various matters ranging from disease control, to garbage collection and hygiene. He saw a direct correlation between the resistance of the people of Guet-Ndar to sanitary and medical measures and the formation of a permanent focus of infection, on the one hand, and the continued spread of plague epidemic in the entire colony, on the other hand. He concluded his message by stating that only the rule of law in its entire rigor could come to terms with such "stubbornness."⁵⁹ Diagne's second intervention failed to convince the leaders of Guet-Ndar to comply with the anti-plague measures.

## The State of Emergency, 13 August–14 November 1918

The strategy of accommodation, referred to by the colonial officials as the "political phase" of the plague policy, ended in early August 1918, when the "mob" chased Mailleraud Frederic, an agent of the Hygiene Service on duty in Guet-Ndar, shouting insults at him. Plague administrators came to the realization that the accommodationist strategy adopted so far was unworkable.⁶⁰ They gave way, and on 13 August 1918, they inaugurated what they called the "medico-military phase" of the plague policy by declaring a state of emergency on Guet-Ndar, Ndar-Toute and on the isolation station of Pointe-aux-Chameaux. The sanitarians were delighted. They then had the opportunity to use the "purifying fire" to destroy *Y. pestis*, as they had hoped since the beginning of the epidemic crisis. Dr. Thoulon, Head of Health Service, claimed that "because of their incurable inertia, the *indigènes* had irritated their opponents and exhausted their best friends."⁶¹

It is interesting to note that not only the medical discourse borrowed the military vocabulary, referring to the people from Guet-Ndar as "rebels" (*habitants réfractaires*), but also the plague operation was to be conducted like a military operation, using military tactics and military personnel. Prophylactic and sanitary measures were secondary. The plague operation included the immobilization of the fishermen's canoes to prevent escape, the occupation of targeted neighborhoods, the progressive evacuation of the people to the lazaretto for 10 days before being sent to two temporary segregation villages in Sor and in the hinterland of Saint-Louis, the systematic disinfection of few houses that were built in durable materials, and the burning down of the majority of suspected huts. The instructions given to the plague administrators were strict and uncompromising:

> It (the operation) is not about shuffling, negotiating, sparing such hut, sparing such notable, (hesitating) instead of burning down: weapons must speak—with cold steel, for the *Tirailleurs* (infantry) will have only the butt and the bayonet, the *Spahis* will only use their sword or the clog of their mount.[62]

The operation started on 14 August 1918 at 4 p.m. after the fishermen returned home, and was met with fierce resistance. Men, women and children were all armed with sticks, *matraques*, iron bars, harpoons, knives, hatches, and *emerillons*. Within three days, given the imposing military force deployed, the resilient residents of Guet-

**Figure 6.1. Evacuations of the Residents of Guet-Ndar, Saint-Louis, 1918**

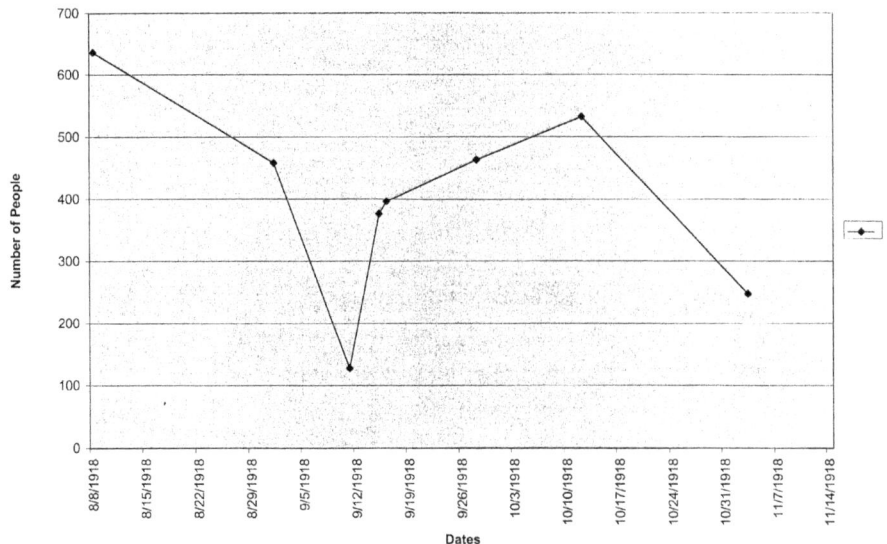

*Source:* ANS/Senegal/H49, Report on the Evacuation of Guet-Ndar (14 Aug.–14 Sept. 1918), by Dr. Thoulon.
a. ANS/AOF/H56/48, *Commandant Militaires* to governor general, Sept. 29, 1918.

Ndar gave up all opposition and decided to negotiate. On 17 August, a letter addressed to Governor Leveque by the "*habitants de* Guet-Ndar" reached the desk of the *Secrétaire Général*. Written in broken French, it is almost incomprehensible because of the form of sentences, grammar and disjointed ideas which reflected the problems the colonial administration faced in making the Africans French. A close reading shows that the anonymous authors of the letter tried to present a counter-argument to all the accusations made against them. They attributed various incidents that had occurred to serious "misunderstandings" and "non-comprehension." They denied being "resistors" and reminded the governor that they had a long tradition of cooperation with the colonial authorities going back to the time of Governor Faidherbe in the 1850s and 1860s. The main evidence presented as the most recent example of cooperation with the administration was the military conscription during the First World War. After setting the record straight, the letter's authors complained about the negative impact of the sanitary *cordon* on their fishing industry and about the non-payment of allocations to the families of the soldiers sent to the front lines. In conclusion, they reiterated their attachment to the *patrie* (France) and to Guet-Ndar, the land of their ancestors, and promised to cooperate with plague administrators in the hope that, after their stay in the isolation camps, they would be permitted to resettle in Guet-Ndar and resume their normal fishing activities.[63] The letter from the residents of Guet-Ndar never reached the governor's desk. It was discarded because of its anonymous character and classified by the hardliners within the administration *sans suite*.[64] Argument could be made that by early August the colonial officials' construction of the plague epidemic was deeply influenced by the doctors who had lobbied all along for a "radical solution" to the Guet-Ndar problem in order to achieve "the final extinction of this dangerous focus" of infection.[65]

An orderly evacuation of the people of Guet-Ndar began on 18 August 1918. Groups of women, children, the elderly and the sick in convoys were transported in canoes, while men, escorted by the *Spahis* (cavalry) walked to the lazaretto at the Pointe-aux-Chameaux on the Langue de Barbarie, as figure 6.1 indicates.

The dwellings at Pointe-aux-Chameaux could accommodate a total of 400 people at one time. The duration of the quarantine was 10 days. In any case, by mid-September the leading protestors surrendered to the security forces after receiving assurances that they would receive special wards in the isolation camp. In early November Guet-Ndar was empty and ready for the "purifying fire" which consumed all but 126 dwellings built in durable materials.[66] The medico-military phase was then replaced by the "medico-hygienic" phase of the plague policy. Governor General Angoulvant estimated the expenses engaged in the operation of evacuation of Guet-Ndar and the destruction of homes at around 1,000,000 francs.[67]

The very operation of the quarantine itself in the lazaretto presented the administration with a logistical challenge that became a real nightmare. Carrera, administrator of the plague policy in Saint-Louis, received many complaints about the inadequacy of food and water supplies, transportation of health officers, nondiscipline of plague administration agents, inexperience of canoeists, and delays in the payment of indemnity to the troops enforcing the quarantine. Abdoulaye Wade, a local notable whose mother, niece, and two sisters—one having three little children—were isolated at the lazaretto at Pointe-aux-Chameaux, accused the administration officials of using a double standard in their isolation practices by making exceptions and allowing some city residents to be isolated in their own homes, while rejecting his request to keep his relatives in his house that was disinfected. His mother and one of his sisters died of plague in the lazaretto. From his perspective, a terrible injustice was done to him and others.[68] Other complaints came from people from out of town who simply found themselves in the wrong place at the wrong time and who wanted to go back to their families.[69] The year 1918 saw a sharp increase in plague morbidity and mortality, with 1,233 officially listed cases and 1,033 officially listed deaths.[70] Upon the completion of the quarantine in the lazaretto, 36 alleged leading resistors were arrested and detained. Later, 18 were released for various reasons.[71]

By early January 1919, the plague operation in Saint-Louis had created serious administrative problems. After a ten-day stay in the lazaretto, people who had nowhere to go were settled in two segregation villages, one in the southern part of Sor and the other in the hinterland of Saint-Louis, with the capacity to accommodate 1,000 people[72] out of 7,000 ex-residents of Guet-Ndar. Others became homeless. A reporter of the newspaper *La Tribune* criticized the colonial authorities about the fate of the people left homeless after the destruction of their homes.[73] Some notables from Saint-Louis, whose relatives lived in Guet-Ndar, complained about the social character of the plague policy because residents of the city-island (the elite and the middle class) were accorded a special treatment; they were isolated in their own homes, not in the isolation camps as was the case for the urban poor.[74] F. Devès, General Councilor, was appalled by the difficult living conditions of the people who were left to fend for themselves in the city streets, including more than 2,000 people camping on the beach in makeshift tents. He blamed the government for the prevalence of "complete anarchy, brutality and illegality" during the post-quarantine period and demanded explanations to the governor general about the crisis.[75]

After a period of vacillation and shuffling and under intense pressure from local elite and Paris, the administration finally decided to find solutions to the crisis. An Evaluation Commission was set up to propose the compensation to the residents of Guet-Ndar for the loss due to the plague operation. In 1919 there was a substantial decline in plague mortality estimated at 257 deaths. The most crucial issue related to the reoccupation of Guet-Ndar remained under investigation until

September 1919 when the authorities agreed to rebuild Guet-Ndar according to a new master plan to be designed, which would have provisions for large spaces and street alignment and give priority to hygiene and waste removal.[76] The decision to rebuild Guet-Ndar can be seen as reflecting the final triumph of the moderates within the administration who opposed the suggestion made by the Hygiene Colonial Committee to create several new ethnic slums instead. In October 1920, the new mayor of Saint-Louis, N'Diaye Cledor Amadou, provided a total amount of 5,000 francs for distribution among the most destitute people of Guet-Ndar.[77] The Commission was presented with fantasist claims. El Hadj Macaty Fall requested compensation for a building that never existed. Others misrepresented the rent they paid before the burnings jumping from 40 frs. to 300 frs. per month. The estimated value of the building John Beigh went from 200 to 1,000 frs. It was evident that everyone wanted to take advantage of the victim compensation fund.[78] As far as mortality for 1920 is concerned, the statistics indicate a total of 197 plague victims. Thus, the cost of extinguishing plague in Saint-Louis, the capital of Senegal and of Mauritania, and of protecting Dakar, capital of French West Africa, from a second plague outbreak left an embittered population.

## Conclusion

The 1917–1918 plague epidemic in Saint-Louis revealed continuity and change in the French medical policy with reference to previous epidemics. The French experience with yellow fever and cholera in Saint-Louis presented many similarities with the 1914 Dakar plague epidemic in that, as Myron Echenberg has suggested concerning Dakar, the French medical policy was authoritarian and dismissive of African concerns.[79] In the nineteenth century, it was the members of civil society who led the battle for a balance between health concerns, commercial interests, and civil liberties. What was different in 1917–1918 was that there was a division within the French community between hardliners (physicians), who were in favor of the forced removal policy, and moderates (governor and governor general), who were willing to take into account financial considerations, civil liberties and some aspects of African culture and religion that were compatible with Western medical and sanitary measures. It took the administration several months of negotiations before opting for the forced evacuation of the people of Guet-Ndar. Governor Lévecque never made the kinds of concessions he made to the Africans in Dakar concerning the vaccination card, the possibility to isolate plague suspects in their own neighborhoods for medical surveillance,[80] or leaving the dead bodies with their relatives for dignified burials; these were the contentious issues that provoked popular resistance to sanitary and medical anti-plague measures. The grave consequences of using armed force against people whom they had classified as unsanitary subjects led the colonial authorities to abandon the "ethnic villages" project in favor of rebuilding Guet-Ndar. This change of mind shows that accommodation became a viable policy choice contributing to political stability.

## Notes

1. Wayne Biddle, *A Field Guide to Germs* (New York: H. Holt, 1995), 166.
2. Ann G. Carmichael, "Bubonic Plague," in *The Cambridge World History of Human Disease*, K. F. Kipple (ed.) (Cambridge; New York: Cambridge U. Press, 1993), 628–30.
3. Myron Echenberg, *Black Death, White Medicine: Bubonic Plague and the Politics of Public Health in Colonial Senegal, 1914–1945* (Portsmouth, NH: Heinemann, 2002), 19.
4. J. Brossollet and H. Mollaret, *Pourquoi la peste? Le rat, la puce et le bubon* (Paris: Gallimard, 1994), 14.
5. J. Brossollet and H. Mollaret, *Pourquoi la peste? Le rat, la puce et le bubon*, 84–97.
6. Myron Echenberg, *Black Death, White Medicine*, 20, 120–123.
7. ANS/Senegal/H30, Hygiene and Public Sanitation Colonial Committee meeting records, May 20, 1914; ANS/Senegal/H30, governor to mayor, July 1914; ANS/Senegal/H30, governor to Health Service, Dec. 13, 1914.
8. ANS/Senegal/H49, Arrêté declarant contaminée de peste la ville de Saint-Louis, (quartier de N'Dar-Toute, Mosque Sidy Tall), December 7, 1917.
9. ANS/Senegal/H76, Police de la ville de Saint-Louis, rapport du 22–24 décembre, 1917.
10. ANS/Senegal/H49, Arrêté interdisant aux habitants de Guet-Ndar et Ndar-Toute pendant toute la durée de l'épidémie, de changer de domicile sans autorisation préalable de l'autorité sanitaire, December 10, 1917.
11. ANS/Senegal/H49, Arrêté interdisant l'accès du quartier de Ndar-Toute contaminé de peste, December 12, 1917.
12. Arrêté nommant une commission chargée de procèder à l'évaluation des constructions et objets divers dont la destruction aura été ordonnée par l'autorité sanitaire, December 11, 1917.
13. ANS/Senegal/H 76, Police report, 15–16 Febr., 14–15 Mars 1918.
14. ANS/Senegal/H76, Rapport de l'Inspecteur de Police, Albertini, Dec. 23, 1917; ANS/Senegal/H76, telegram of Governor General Angulvant to governor, March 16, 1918.
15. ANS/Senegal/H76, Note de service de Muller au gouverner, Dec. 24, 1917.
16. ANS/AOF/H57/31, 29-30. The penalties for non-compliance with sanitary and medical measures were either too severe (ordinance of Dec. 12, 1912) or too lenient, weak and inoperative (decree of April 14, 1904 related to the protection of public health in French West Africa) to be imposed by the courts.
17. ANS/Senegal/H76, Official telegram of interim *Secrétaire Général* Muller to lieutenant-governor Bakel-Maka-Colybentan, no. 409 of Jan. 16, 1918.
18. ANS/Senegal/H76, Inspection of Sanitary Services to governor general, Jan. 19, 1918.
19. ANS/Senegal/H76, "Habitants of Guet-Ndar" to governor, Jan. 18, 1918.
20. ANS/Senegal/H76, Police report, Jan. 22–23, 1918; also ANS/Senegal/H76, Dr. Gallen to Head of Medical Service, Jan. 22, 1918.

21. ANS/Senegal/H76, telegram of governor to governor general, no. C44 of Jan. 17, 1918.
22. ANS/Senegal/H76, Police report, 1–2 March 1918.
23. ANS/Senegal/H49, Chief Medical Officer to Lieutenant-Governor of Senegal, January 24, 1918, no. 31ss.
24. ANS/Senegal/H49, handwritten note, n.d., n.a. provided.
25. ANS/Senegal/H49, Dr. Damian to Inspector of Sanitary and Health Services, no. 58 S.S. of February 11, 1918.
26. ANS/Senegal/H49, telegram of *Commissaire de République*, Diagne, to Mayor of Saint-Louis, February 21, 1918, no. 10.
27. ANS/AOF/3G3–7/38, Requisition, by Lévecque, Febr. 28, 1918.
28. Arrêté édictant des mesures sanitaires pour la ville de Saint-Louis, pendant la durée de l'épidémie de peste, February 28, 1918, *Journal Officiel du Sénégal*, February 1918.
29. ANS/AOF/H49, Cordon sanitaire. Consignes pour les Chefs de poste, March 2, 1918.
30. ANS/Senegal/H49, governor to governor general, March 15, 1918.
31. ANS/Senegal/H76, Angoulvant's telegram to Lévecque, March 16, 1918.
32. The Hygiene Colonial Committee included Dr. Contaut (Head of Health Service), Dr. Fulconis (Colonial Ambulance), Dr. Basque (Colonial Ambulance), Dr. Dupont (municipal Hygiene Service Head), Mayor Chimère, public works director Grimaux, veterinarian Teppaz, and three notable residents (Douan, Cales, and Marsan).
33. ANS/Senegal/H49, minutes of the Colonial Hygiene Committee, March 21, 1918.
34. ANS/Senegal/H 49, Colonial Hygiene Committee meeting minutes, March 21, 1918.
35. ANS/Senegal/H49, Dr. Contaut to the Lieutenant-Governor of Senegal, no. 108 L.S. of March 22, 1918.
36. ANS/Senegal/H49, Autopsy report done by Dr. Fulconis, March 22, 1918.
37. ANS/Senegal/H49, Dr. Contaut to the Lieutenant-Governor of Senegal, no. 108 L.S. of March 22, 1918.
38. ANS/Senegal/H49, Dr. Contaut to the Lieutenant-Governor of Senegal, no. 108 L.S. of March 22, 1918.
39. ANS/Senegal/H49, Governor Lévecque to governor general of FWA, March 25, 1918.
40. ANS/Senegal/H49, Governor Lévecque to governor general of FWA, March 25, 1918.
41. ANS/Senegal/H49, Governor Lévecque to governor general of FWA, March 25, 1918.
42. ANS/Senegal/H74, notables from Guet-Ndar to lieutenant-governor, March 1, 1918.
43. ANS/Senegal/H74, telegram of cabinet director, Muller, to governor in mission in Ziguinchor, March 7, 1918.
44. ANS/Senegal/H74, "Habitants de Guet-NDAR" to François Carpot, March 7, 1918.
45. ANS/Senegal/H49, Police report, March 26, 1918.
46. ANS/Senegal/H49, Special report on plague in Saint-Louis (15–31 March), April 3, 1918.
47. ANS/Senegal/H49, Official telegram, Governor General Angoulvant to governor of Senegal, no. 365 of April 3, 1918.
48. ANS/Senegal/H49, Police report, n.d. mentioned.
49. ANS/Senegal/H49, Special report on plague in Saint-Louis (1–15 April 1918), by Dr. Contaut, April 17, 1918.
50. ANS/Senegal/H49, governor general of FWA to lieutenant-governor of Senegal, April 22, 1918; ANS/Senegal/H57/31, 31.
51. ANS/AOF/H57/31, 32.

52. ANS/Senegal/H49, governor general to president of General Council, May 4, 1918.
53. ANS/Senegal/H77, Extrait des délibérations de la Commission Coloniale, Jan. 14, 1918.
54. ANS/Senegal/h 49, governor general of FWA to president of *Conseil Général*, May 4, 1918.
55. ANS/AOF/H56/117, Arrêté déterminant les mesures propres à enrayer l'épidémie de la peste constatée dans certaines régions du Sénégal, June 25, 1918; ANS/AOF/H56/117–6, Arrêté no. 806 déclarant contaminées de peste la ville de Saint-Louis et les escales de la voie ferrée Dakar-Saint-Louis, et en danger de contamination la ville de Dakar, et déterminant les conditions dans lesquelles seront faites obligatoirement la vaccination et la re-vaccination dans ces agglomérations urbaines et suburbaines.
56. ANS/AOF/H57/31, 33.
57. ANS, *Moniteur du Sénégal*, 413.
58. ANS/AOF/H57/31, 34–5.
59. ANS/Senegal/H77, Angoulvant to Commissaire de la République, no. 59 of Aug. 1, 1918.
60. ANS/Senegal/H49, agent Mailleraud Frederic to Hygiene Service physician, August 9, 1918.
61. ANS/Senegal/H49, Report on the evacuation of Guet-Ndar (14 Aug.–14 Sept. 1920), by Dr. Thoulon.
62. ANS/Senegal/H49, Dr. Thoulon.
63. ANS/Senegal/H49, Residents of Guet-Ndar to lieutenant-governor of Senegal, August 17, 1918.
64. ANS/Senegal/H49, cabinet's minute, March 11, 1918.
65. ANS/AOF/H57/31, 35.
66. ANS/Senegal/H73, États des baraques épargnées à Guet-Ndar 1920.
67. ANS/AOF/H57/31, 32.
68. ANS/Senegal/H74, Abdulaye Wade to governor, Dec. 6, 1917.
69. ANS/Senegal/H74, unclassified document.
70. ANS/AOF/H57/18.
71. ANS/AOF/H56/51, lieutenant-colonel Lafitte to *Commandant Supérieur des* Troupes, Sept. 27, 1918.
72. ANS/AOF/H57/31, 33.
73. ANS/AOF/P165/90, governor general to governor, January 13, 1919 about the rebuilding of Guet-Ndar. See also *Tribune*, no. 66, January 5, 1919, 1.
74. ANS/Senegal/H74, Abdulaye Wade to governor, December 6, 1918.
75. ANS/AOF/3G3.7/202, F. Devès to governor general, February 23, 1919; ANS/AOF/3G3.7/210, letter of March 5, 1919.
76. ANS/AOF/3G3.7/218, governor general to governor, no. 1196 of September 10, 1918.
77. ANS/Senegal/H78, Ordinance of mayor N'Diaye Cledor Amadou, Oct. 25, 1920.
78. ANS/Senegal/H78, Ordinance of mayor of Saint-Louis, Oct. 25, 1920.
79. Myron Echenberg, *Black Death, White Medicine*, 242.
80. Myron Echenberg, *Black Death, White Medicine*, 186.

# Conclusion

This book has focused on the challenges the French public health authorities faced to make Senegal a healthy colony and Saint-Louis, the capital of Senegal, of Mauritania, and, for a short period, of French West Africa, the healthiest capital/port-city during the formative years of the colonial state. The emergence and re-emergence of epidemic diseases frustrated the priorities of the colonial authorities. The constructions of yellow fever, cholera and plague epidemics were informed not only by the miasmatic theory, which dominated the medical thinking until early 1901, but also by the perceptions the colonists had of African societies, cultures, and environment. In the absence of effective therapies, these perceptions and contemporary medical knowledge and practices help explain why state resources and health policies were directed to one direction rather than another. There were also differences in perceptions and responses between one epidemic disease and another. Although Saint-Louis' health problems were due to the importation of yellow fever, cholera, and plague on the one hand, and to the inadequacies of public health policies concerning housing, waste removal, and clean water supply on the other, the colonial, medical and missionary discourses constructed them differently. The public health authorities viewed yellow fever as the "White man's disease" and set up their priorities to protect the health of the European military and populace at large. Medical and missionary discourses constructed cholera as a "Black man's disease" but for totally different reasons. Plague epidemic that occurred well into the colonial moment dominated by the germ theory revealed the continuity of the same discourses and practices.

Epidemic diseases consolidated the professional and social status of the physician within the colonial bureaucracy and made him a key player in the decision-making process affecting all colonial matters without distinction. The threat of disease provided an impetus to the creation and expansion of the Health Service of Senegal and various health boards controlled by the physicians. By the 1890s the Chief Medical Officer was a member of the *Conseil d'Administration* and *Conseil Privé*. The administrative problems linked to the management of epidemic diseases underlined the impact of political constraints on medical and sanitary intervention. The rationality and agenda of the administrative authorities were different from and in competition with those of the physicians and the merchants.

The control of epidemic diseases brought to the fore the questions that were left unresolved about the place of the African urban poor in the colonial city, about their stigmatization and the issues of cohabitation or segregation due to the perception that they were carriers of infectious agents, and about the very tropicality of Senegal. The outbreak of bubonic plague epidemic in 1917 led to the adoption of unprecedented sanitary and medical measures, which provoked violent reactions among the urban poor. Guet-Ndar, the most populous slum of Saint-Louis that was seen as a "space of death," stood as the symbol of everything that was wrong with French Senegal. On the issue of scapegoating the victims, Saint-Louis was not alone. Similar processes of marginalization had accompanied, in the late nineteenth- and early twentieth centuries, the outbreaks of bubonic plague in Cape Town and in Dakar, tuberculosis in South Africa, and cholera in India and Brazil, and smallpox, plague, and syphilis in San Francisco. In these cases, like in Saint-Louis, the official declaration of the epidemic disease, which did not correspond with its real appearance, was also followed by panic and the initial phase of disorientation, then came the flight and the search for scapegoats: the poor, prostitutes, ethnic minorities, religious or cultural beliefs and practices, or lifestyles. Thus, epidemic diseases revealed the urban society to itself: unequal development, prejudices, heterogeneity and tensions across neighborhoods, social classes, religions, races and ethnic groups, but also, to some extent, homogeneity and solidarity within them. The bubonic plague epidemic showed the continuity of earlier practices despite the fact that the bacteriological revolution provided a better understanding of the etiology of diseases. Where the new knowledge brought change was in the area of vector control with an effective sanitation campaign—including the construction of docks, the drainage of the swamps, and the creation of the Hygiene Service—and water purification. The irony is that, when Saint-Louis' health problems were finally seriously addressed, the city lost its preeminence to Dakar, which became the capital of French West Africa in 1902 and where the institutions were transferred in 1904.

One of the long-lasting legacies of disease in colonial Senegal has been the reproduction of the colonial mentality in the postcolonial state of Senegal, including the tendency to search for scapegoats whenever medical events overwhelmed the capacity of existing infrastructure and resources. Residents of Guet-Ndar continued to be viewed as "rebels," who do not obey any laws. The advent of HIV/AIDS in the 1980s posed a new challenge to state officials, who, instead of understanding how poverty affected the spread of the virus, were tempted to blame the sufferers' cultural and religious practices.

# Sources

## Archival Sources

### A. 1. Archives Nationales de la France, Section Outre-Mer (Aix-en-Provence; ANF/SOM)

Dépôt des Fortifications. Sénégal, 1670–1889. Séries Géographiques
Sénégal et Dépendances VII.Conseil d'Administration et Conseil Privé
Dossiers 22-24: Conseil d'Administration, 1822–1900. Dossiers 25–26: Conseil Privé, 1822–1900.
Sénégal et Dépendances XI. Police, Hygiène, Assistance.
Dossiers 1–26: Police. Passagers. Prisons. Voirie. Innondations. Décorations, 1815–1895.
Dossiers 27–42: Service Sanitaire. Hôpital. Personnel Médical. Epidémies. Lazarets, 1815–1895.
Dossiers 43–46: Assistance Publique: Hospices, Secours votés par le Conseil Général, 1816–1895.
Dossiers 47–51: 1895–1910. Sénégal et Dépendances XII. Travaux Publics.
Dossiers 7–10; 17–21: Travaux Publics. Guinée IV. Expansion Territoriale, 1889–1895.
Dossier 1, b: Relations avec le roi Dinah Salifou.
Dossier 6, g: Politique Indigène, Dinah Salifou. Agence Economique de la France d'Outre-Mer. Photographies.

### A. 2. Archives du Service Historique de la Marine (Château de Vincennes, Paris).

Archives Centrales de la Marine.
Dossier CC7: Dossiers Personnels des Officiers.

## A. 3. Archives Nationales du Sénégal (Dakar; ANS)

### SÉRIE H: SANTÉ (A.O.F.).

H 1 à H 26: Organisation et Fonctionnement (1829–1920).

H 1. Rapports médicaux des divers postes de la colonie adressés au médecin en chef du Service de Santé (1845–1865, 1880–1883); N'Dar-Toute (1850–1).

H 2. Correspondance départ du Service de Santé adressée au Gouverneur, aux Chefs de Service, etc. (12 avril 1859–1 mai 1875).

H 3. Copies des dépêches du Ministre et du Directeur des Colonies adressées aux Gouverneurs et relatives au Service de Santé (14 mars 1851–5 mai 1857).

H 6. Santé. Organisation et fonctionnement, 1829–1846—Rapport sur le Service de Santé de la colonie (1829), épidémies (1830).

H 7. Santé. Organisation et fonctionnement—Procès-verbaux de la commission chargée de l'oeuvre de l'assistance aux populations malheureuses (1865–6), quarantaines, hôpitaux, patentes de santé, analyse des eaux, dissémination...

H 9. Santé. Organisation et fonctionnement, 1895–7—Pièces périodiques, correspondance, aliénés.

H 10. Santé. Organisation et fonctionnement, 1895–1902. Etat sanitaire et mesures de quarantaine; Sénégal; Guinée; Côte- d'Ivoire.

H 12. Santé. Organisation et fonctionnement en A.O.F., 1902–1911. Police sanitaire maritime, projet de maternité, conseils de santé, rapports médicaux, organisation sanitaire, assistance indigène, hôpital de Bamako...

H 13. Santé. Organisation et fonctionnement en A.O.P., 1912–3.

H 14. Santé. Organisation et fonctionnement en A.O.F., 1914–1919... Ecole de Médecine (1918).

H 15. Hydrologie, analyse des eaux, 1896–1919.

H 16. Dix ans de nosologie à l'hôpital de SL par le docteur Carpot (1889–1899). 1 notice.H 17. Dissémination des troupes et rapatriement du personnel en cas d'épidémie. Etudes, notes et instructions, 1902–3.

H 18. Campagnes contre les moustiques et assainissement de Gorée et Dakar. Etudes, rapports, réglementations, 1904–5.

H 19. Comité Supérieur et Services Municipaux d'Hygiène. Organisation et fonctionnement, 1904–1910.

H 25. Oeuvre d'assistance aux enfants métis. Pensionnat d'enfants métis; assistance au Haut-Sénégal-Niger; condition juridique des métis, 1916–1918.

H 26. Aliénés, 1916–8.

H 27 à H 57: Epidémies (1868–1920)

H 27. Choléra. Saint-Louis, Bakel, etc., 1868–1869.

H 28. Fièvre Jaune. 1. Correspondence reçue par le Gouverneur.

H 29. Fièvre Jaune. 2. Listes et avis de décès, 1878–79.

H 32. Fièvre Jaune. Saint-Louis, 1878–82.

H 36. Choléra. 2 Saint-Louis, 1893.

H 37. Variole. Saint-Louis, 1886-94.

H 38. Variole. Saint-Louis, 1886–1909.

H 39. Epidémies et maladies contagieuses, 1895–1911.

H 41. Fièvre Jaune. 1. Mesures sanitaires et quarantaines, 1900.

H 42. Fièvre Jaune. 2. Cordons sanitaires et incidents, 1900.
H 44. Fièvre Jaune. 4. Textes réglementaires et questions diverses, 1900. H 45. Fièvre Jaune. 5. Bulletins sanitaires et avis de décès.
H 48. Mission sanitaire du Sénégal en 1901.
H 49. Mission sanitaire du Sénégal. Correspondances, 1901.
H 50. Fièvre Jaune. A.O.F., 1901.
H 52. Fièvre Jaune. Sénégal . . . , 1903.
H 54. Fièvre Jaune. Mesures préventives. Plan d'évacuation, toile métallique, campagne prophylactique, réclamations, 1910–13.
H 56. Peste à Dakar et au Sénégal, 1917–20.

## SERIE H: SANTE (SENEGAL)

H 1. Correspondence du Chef du Service de Santé reçue par le Gouverneur, 1851–1863; lettres de Gorée.
H 2. Correspondance du Gouverneur, 1878–79; Régime sanitaire au Sénégal, 29/8/1884; rapatriement des indigents, 1882–1901.
H 4. Correspondance adressée au Directeur de l'Intérieur par le Gouverneur, 1890–1892.
H 5. Correspondance adressée au Gouverneur du Sénégal par le Chef de Service de Santé, 1899–1906.
H 6. Correspondance du Gouverneur 1902-1903; demandes de rapatriement, 1905–1906.
H 7. Correspondance administratif au Gouverneur par le Service Santé, 1904–1908.
H 8. Correspondance du Gouverneur, 1904; correspondance reçue du Ministère des Colonies, 1904–1906; fièvre jaune: commandes de matériels, 1911–12.
H 9. Correspondance: Directeur du Laboratoire de bactériologie de Saint-Louis, Mai–Décembre 1905; médecin du Service de d'Hygiène Mai–Novembre 1905; Chef de Service de Santé du Sénégal au Gouverneur, 1918–1919; billets d'hospitalisation, 1905.
H 10. Incidents lors du passage du Paraguay, le Octobre–Novembre; Correspondance du Chef de Service de Santé de la colonie, 1906–1915.
H 11. Correspondance administratif au Gouverneur par le Chef de Service de Santé, 1913–34.
H 12. Correspondance administratif au Gouverneur, 1913–6.
H 13. Correspondance du Gouverneur du Sénégal sur la situation sanitaire, 1916–7.
H 15. Décisions du Gouverneur du S 9/8–23/12/1918–19.
H 16. Décisions du Gouverneur du Sénégal concernant le personnel de la Santé, 1914–15; mesures d'hygiène, 1916.
H 18. Régime sanitaire, 1892–1907.
H 19. Régime sanitaire, 1912; personnel des services sanitaires, 1912–14.
H 20. Hygiène et salubrité publique, 1858–78.
H 21. Hygiène. Saint-Louis, 1864–77.
H 22. Régime sanitaire: hygiène et salubrité publique, 1873–1901.
H 23. Hygiène: décrets et arrêtés, 1897–1912.
H 24. Hygiène: arrêtés sur la police sanitaire maritime dans les colonies et les pays de protectorat, 1899.
H 25. Conseil sanitaire: procès-verbaux des séances, 1900–1.
H 41. Décret du 7 août 1885 réorganisation du Service de Santé de la Marine.

H 42. Fièvre Jaune: mise en quarantaine des provenances de Dakar, 1874–1882.
H 43. Fièvre Jaune: région du fleuve: mise en quarantaine, 1878.
H 44. Variole: mesures sanitaires, 1881-95. Saint-Louis, 1891–1914.
H 45. Assistance publique: admission des malades mentaux à l'hospice civil, 1896–98. H 46. Régime sanitaire, 1892–97.

### SERIE P: TRAVAUX PUBLICS
P 111. Bâtiments de Saint-Louis. Hôpitaux civil et militaire, 1830–61.
P 165. Assainissement de Saint-Louis, 1898–1912.
P 166. Assainissement de Saint-Louis, egouts, dossier d'adjudication, 1910.

## A. 4. Archives Municipales de Saint-Louis (Hôtel de Ville).

1 A 63. Décrets et circulaires, 1882–1959.
1 B 61. Diverses correspondances, 1886–1959.
1 F 13. Gestion, 1895.
1 K 53. Rapports, 1894-1927; électricité, 1913–1928.
1 O 41. Lotissement, 1852–1947; barrage de Khor, 1904.
1 O 44. Procès-verbaux des tombes à Sor, 1913.
1 O 95. Terrains communaux, 1888–1946.
Q 80. Recensement général de la population, 1914.
B 4. Correspondance, 1917–1927.
2 B 6. Correspondance, 1907–1916.
2 B 20. Correspondance, 1881–1899.
2 B 24. Correspondance, 1909.
3 B 25. Correspondance, 1915.
2 B 48. Correspondances au Maire, 1909–1939.
2 B 97. Diverses correspondances, 1896–1928.
2 B 111. Diverses correspondances, 1891–1956.
2 O 78. Cimetière de Sor, cimetière musulman, 1862–1944.
2 P 55. Eaux et électricité, 1894–1936.
2 P 58. Electricité, 1910–1936.
2 P 88. Eclairage, 1887–1932.
P 59. Electricité, 1913, 1928.
A 60. Décisions et circulaires, 1912, 1959.
3 F 12. Dépenses, 1912.
6 F 49. Electricité, 1910.
E 22. Procès-verbaux du Conseil Municipal, 1910–1946.

## A.5 Missionary Archives

Sisters of Cluny (Paris)
Bulletin de la Congrégation de Saint-Joseph de Cluny, 1885–1904, Correspondence.
Fathers of *Saint-Esprit*.

*Journal de la Communauté de Saint-Louis*, 1852–1890, 1891–1904, 1904–1951.
*Echo de Saint-Louis*, 1906–1920.
*Bulletin de la Congrégation*, 1856–1902.
General Correspondence
boite no. 157–B on Senegambia.
boite no. 158–B on Senegal.

## Oral Sources. Interviews in Saint-Louis

Mme André at Sor (Corniche) on 16 September 1994.
G. Bonnet Georgette at Sor (Corniche) on 15 September 1994.
Moctar Diagne at N'Dar Toute on 16 April 1994.
Ahmat Diba at Cité Niakh on 24 June 1994.
Cheikh Diouf at Cité Niakh on 20–25 June 1994.
Pape Cey Charles at Sor on 16 April 1994.
Maguette Cey at Sor on 23 April 1994.
Mile Diop at the Université de Saint-Louis on 18 January 1995.
Marieme Gueye at Guet-N'Dar on 26 April 1994.
Labelle Louise (née Carrère) at Sor on 17 September 1994.
Léon Diakhate at Khor on 23 January 1995.
Gabrielle Dansokho at Khor on 23 January 1995.
Mansa Kissi Kamara at Sor on 20 January 1995.
Abdoul Wahab Macodé Ndiaye at Sor on 22 April 1994.
Mbacke Ndiaye at Sor (SO.N.E.E.S.) on 18 May 1994.
Louis Negre at quartier commercial on 16 September 1994.
Prof. Niane (Univ. St.-Louis) at Sor on 7 January 1995.
Mme Niang at the Université de Saint-Louis on 23 April 1994.
Ndoye Samba at Sor (Corniche) on 14 September 1994.
Omar Sarr at quartier sud on 7 April 1994.
Caty Diop Sidi at Sor on 22 April 1994.
Ibrahima Tabane at N'Diolofene on 27 January 1995.
Abubacar Sidiki Tounkara at Guet-N'Dar on 20 January 1995.
Fara Penda Yade at *quartier nord* on 13 June 1994.

## Unpublished Dissertations

Bancal, "Des fièvres au Sénégal," Doctoral thesis in medicine, Montpellier, 1834.
Baril, C., "Souvenirs d'une expédition militaire au Sénégal pendant l'Epidémie de fièvre jaune de 1878," Doctoral thesis in medicine Montpellier, 1883.
Barrows, L. C., "General Faidherbe, the Maurel et Prom Company and French expansion in Senegal," Ph.D. thesis, U. of California, Los Angeles, 1974.

Bax, J. B., "Considérations hygiéniques médicales sur la colonie du Sénégal et sa garnison," Doctoral thesis in medicine, 1830.

Berger, C.-V., "Considérations hygiéniques sur le bataillon des tirailleurs sénégalais," Doctoral thesis in medicine, Montpellier, 1868.

Berville, N., "Remarques sur les maladies du Sénégal," Doctoral thesis in medicine, Paris, 1857.

Bonus, A., "Quelques considérations médicales sur le poste de Dagana, Sénégal," Doctoral thesis in medicine, Montpellier, 1864.

Cheve, B., "La fièvre jaune au Sénégal," Doctoral thesis in medicine, Paris, 1836.

Conklin, Alice Louis, "A Mission to Civilize: Ideology and Imperialism in French West Africa, 1895–1930," Ph.D. dissertation, Princeton University, 1989.

Conley Barrows, Leland C., "General Faidherbe, the Maurel and Prom Company, and French Expansion ni Sénégal," Ph.D. thesis, U. of California, Los Angeles, 1974.

Diop, Angélique, "Santé et colonisation au Sénégal, 1895–1914," Doctoral thesis, 3e cycle, Paris I, UER 09 History, CRA, 1982.

Gille, Bernard, "Les Services d'Hygiène du Sénégal de 1905 à 1920," Doctoral thesis, 3e cycle, Université de Provence, Institut d'Histoire des Pays d'Outre-Mer, Aix-en—Provence, 1974, 2 vols.

Grimaud, Aimée (born Houemavo), "Les Médecins Africains en A.O.F.: Etude socio-historique sur la formation d'une elite coloniale," Mémoire de Maitrise, Université de Dakar, 1979.

Hilary, J., "French Citizens or Imperial Agents? The Métis of Saint-Louis and Republicanism in the Colonial Period, 1971–1920," Ph.D. dissertation, Michigan State U., 2003.

Kouyate Mamadou Moustapha Dieng, "Famines et disettes et epidémies dans la basse et moyenne vallée du fleuve Sénégal, 1854–1945," Doctoral thesis, 3e cycle, Université Cheikh Anta Diop de Dakar, 1992.

Kouyate, Seydou (Dr.), "Les traitants Africains de la fièvre jaune," Doctoral thesis in medicine, Université de Montpellier, 1955.

Marcson, M., "European-African Interaction in thé Precolonial Period: Saint-Louis, Sénégal, 1758–1854," Ph.D. thesis, Princeton U., 1976.

McLane, Margaret O., "Economic Expansionism and the Shape of Empire; French Enterprise in West Africa, 1850–1914," Ph.D. dissertation, U. of Wisconsin–Madison, 1994.

Pasquier, R., "Le Sénégal au milieu du XIXe siècle: la crise économique et sociale", 5 Vols., Doctoral thesis in History, U. de Paris; Sorbonne, 1987.

Domergue-CloArec (Danielle), "Politique coloniale française et réalites coloniales: l'exemple de la santé en Cote-d'Ivoire, 1905–1958," Thèse doctorat es Lettres et sciences humaines, 1984.

Reyss, Nathalie, "La santé à Saint-Louis du Sénégal à l'epoque pré-coloniale d'après les récits de voyage et romans: alimentation, hygiène, et santé ou le métissage moyen de survie," Doctoral thesis, U. de Paris I, 1981.

Salleras, Bruno, "La peste en 1914: Médina ou les enjeux d'une politique sanitaire," Doctoral thesis, 3e cycle, EHESS, Paris, 1984.

Séné, Moustapha, "Epidémies et politiques sanitaires au Sénégal, 1920–1960. exemples de la fièvre jaune et de la maladie du sommeil (trypanosomiase)," Mémoire de Maitrise, Université Cheikh Anta Diop de Dakar, 1991.

Vauvray, À., "Des accidents cholériques vulgairement appelés N'Diank au Sénégal," Doctoral thesis in medicine, Montpellier, 1866.

Vincent, J., "La fièvre jaune (épidémie de 1878 et 1881 au Sénégal)," Doctoral thesis in medicine, Montpellier, 1886.

## Published Sources: Official and Semi-Official Publications

*Annuaire du Senegal et Dépendances*, 1855–1902.
*Annuaire de la Marine et des Colonies*, 1857–1894.
*Moniteur du Senegal et Dépendances*, 1856 (later *Journal Officiel du Senegal et Dépendances*).
*Revue Coloniale* (later *Revue Algérienne et Coloniale, Revue Maritime et Coloniale*).
*Bulletin du Comité d'Etudes Historiques et Scientifique de l'Afrique Occidentale Française*, 1917–1920.

## Books and Articles

Abbatucci (Dr.), "Le milieu africain considéré au point de vue de ses effets sur le système nerveux de l'Européen," *ANN. HYG. COL.*, t 13 (1910), pp. 328–335.
Alfoldy, Geza, *The Social History of Rome*. Translated by David Braund and Frank Pollock (Baltimore: Johns Hopkins U. Press, 1991).
Alquier, P. "Saint-Louis du Sénégal sous la Révolution et l'Empire, 1789–1809," *B.C.E.H.S. de l'A.O.F.* (1922).
Anonymous, *Une grande missionnaire Anne-Marie Javouhey, 1779–1851* (Osny, 1990).
Arnold, David (éd.), *Imperial Medicine and Indigenous Societies* (Manchester, England: Manchester U. Press, 1988).
Arnold, David, *Colonizing the Body. State Medicine and Epidemic Disease* (Berkeley: U. of California Press, 1993).
Arnold, David, *The Tropics and the Traveling Gaze: India, Landscape, and Science, 1800–1856* (Seattle: University of Washington Press, 2006).
Astrie (Max), "La Nélaouane (maladie du sommeil au S)," *Bull. de la société de geogr de Marseille*, t. XIII, 1889, pp. 29–36.
Aubert, E., "Un peu d' hygiene coloniale," *Les Annales coloniales*, no. 11 (Nov. 1, 1901), 92–93.
Aude, P. (Dr.), *Code des Officiers du Corps de Santé de la Marine* (Paris: Berger-Levrault et Cie, 1877).
Baldwin, Peter, *Contagion and the State in Europe, 1830–1930* (New York: Cambridge University Press, 1999).
Barrows, L. C., "L'Oeuvre, la carière du général Faidherbe et les débuts de l'Afrique noire française: une analyse critique contemporaine," *Le Mois en Afrique* 235–236 (Aug.–Sept. 1985) and 237–238 (Oct.–Nov. 1985).
Barrows, L. C. "The Merchants and General Faidherbe: Aspects of French Expansion in Senegal in the 1850's," *Revue Française d'Histoire d'Outre-Mer*, 223 (1974), 218–234.
Barry, Boubacar, *Le royaume du Waalo: Le Sénégal avant la conquête* (Paris: François Maspero, 1972).
Beck, Ann, *History of the British Medical Administration of East Africa, 1900–1950* (Cambridge, MA: Harvard University Press, 1970).

Bérenger-Féraud (Dr.), "Etudes sur la Sénégambie," *Moniteur du Sénégal et Dépendances*, 1873, pp. 9–10, 19–20, 22–24, 109–112, 126–128.

Bérenger-Féraut, *De la fièvre jaune au Sénégal* (Paris, 1874).

Bérenger-Féraut, *Traité clinique des maladies des Européens au Sénégal* (Paris-La Haye, 1875–1878), 2 vols.

Biddle, W., *A Field Guide to Germs* (New York: H. Holt, 1995).

Biondi, J. P., *Saint-Louis du Sénégal, mémoires d'un métissage* (Paris: Denoel, 1987).

Blalock Jr., Hubert M., *Race and Ethnie Relations* (Englewood Cliffs, NJ: Prentice Hall, 1982).

Bloom, Khaled, *The Mississippi Valley's Great Yellow Fever Epidemic of 1878* (Baton Rouge: Louisiana State University Press, 1993), 364.

Boilat, D., *Esquisses sénégalaises* (Paris: Karthala, 1984).

Bonnardel, R., *Saint-Louis du Sénégal: mort ou naissance?* (Paris: L'Harmattan, 1992).

Bonus, Dr., *Nouvelles recherches sur le climat du Sénégal* (Paris: Annales du Bureau Central Météorologique, 1879).

Bonus, Dr., *Topographie médicale du Sénégal* (Paris: Archives de Médecine Navale, 1880–1889).

Bonus, Dr., *Recherches sur le climat du Sénégal* (Paris, 1875).

Bouche, D. "L'école française du Sénégal de 1850 à 1920," *Revue Française d'Histoire d'Outre-Mer* 223 (1974), 218–234.

Bouche, D. "Les villages de liberté en A.O.F.," Bulletin de l'I.F.A.N., vol. XI, B no. 3–4 (1949), 491–540, and vol. XII, 1 (1950), 135–215.

Bouquillon, Y., and Cornevin, R., *David Boilat (1814–1901). Le précurseur* (Dakar: Nouvelles Editions Africaines, 1981).

Bourdieu, P., *Distinction: A Social Critique of the Judgment of Taste*. Trans. Richard Nice. (Cambridge, MA: Harvard U. Press, 1984).

Brigaud, F., and J. Vast, *Saint-Louis du Sénégal, ville aux mille visages* (Dakar: Édit. Clairafrique, 1987).

Brau, Paul, *Trois siècles de médecine coloniale française* (Paris: Vigot Frères éd., 1931). Exposition Coloniale Internationale de Paris de 1951. Les années françaises d'Outre-Mer. *Le Service de Santé aux colonies* (Paris: Impr A. Lahure, MDCCCCXXXI).

Brenner, Louis, "The 'Esoteric sciences' in West African Islam," in Brian M. du Toit and Ismail Abdalla (eds.), *African Healing Strategies* (New York: Trado-Medic Books, 1985), p. 22.

Brigaud, F., and Vaast, Jean, *Saint-Louis du Sénégal. Ville aux mille visages* (Dakar: Ed. Clairafrique, 1987).

Briggs, Charles L., and C. Mantini-Briggs, *Stories in the Time of Cholera. Racial Profiling during a Medical Nightmare* (Berkeley: University of California Press, 2003).

Brooks Jr., G. E., "The *Signares* of Saint-Louis and Goree: Women Entrepreneurs in Eighteenth-Century Sénégal," in Hafkin, N. J. and Bay, E. G. (eds.), *Women in Africa. Studies in Social and Economic Change* (Stanford, CA: Sanford U. Press, 1976), pp. 19–44.

Brossollet, J., and H. Mollaret, *Pourquoi la peste? Le rat, la puce et le bubon* (Paris: Gallimard, 1994).

Burke, Peter, *The French Historical Revolution: The Annales School, 1929–89* (Stanford, CA: Stanford University Press, 1990).

Bynum, W. E., *Science and the Practice of Medicine in the Nineteenth Century* (New York: Cambridge University Press, 1996).

Camara, C., *Saint-Louis-du-Sénégal. Evolution d'une ville en milieu africain* (Dakar: IFAN, 1968).

Carmichael, Ann G., "Bubonic Plague," in *The Cambridge World History of Human Disease*, K. F. Kipple (ed.) (Cambridge; New York: Cambridge U. Press, 1993), 628–30.
Carpot, Charles, *La fièvre jaune. Epidémie de l'année 1900 à Saint-Louis du Sénégal* (Bordeaux: Impr G. Gounouilhou, 1901).
Carpot, Charles, *La fièvre jaune. Epidémie de l'année 1900 à Saint-Louis. Observations-traitement* (Saint-Louis: Imprimerie O. Lesgourgues, 1912).
Carr, Edward Hallet, *What is History?* (New York: Vantage Books, 1961).
Carre, Adrien, "Hitorique du service de santé de la Marine, 1870–1970", extr de *La Revue Historique de l'armée*, no 1, 1972, pp. 122–156.
Carrère, F., and Holle, P., *De la Sénégambie française* (Paris: Librairie de Firmin Didot Frères, 1855).
Carrigan, Jo Ann, *The Saffron Scourge: A History of Yellow Fever in Louisiana, 1796–1905* (Lafayette: University of Southwestern Louisiana, Center for Louisiana Studies, 1994).
Certeau, Michel de, *La culture au pluriel* (Paris: Seuil, 1974; 1993).
Certeau, Michel de, *L'ecriture de l'histoire* (Paris: Gallimard, 1975; 1984).
Certeau, Michel de, *La fable mystique*, vol. 1: *XVIe-XVIIe siècle* (Paris: Gallimard, 1982; 1987).
Certeau, Michel de, *L'Invention du quotidien*, vol. 1: *Arts de faire* (Paris: Gallimard, 1980; 1990).
Certeau, Michel de, *L'invention du quotidien*, vol. 2: *Habiter, cuisiner*. Avec la collaboration de Luce Giard et Pierre Mayor (Paris: Gallimard, 1994).
Certeau, Michel de, *La possession de Loudoun* (Paris: Gallimard, 1970; Julliard, 1990)
Chevalier, Louis, *Classes laborieuses et classes dangereuses à Paris pendant la Première moitié du XIXe Siècle* (Paris: Plon, 1958).
Christaller, W., *Central Places in Southern Germany*, trans. C. W. Baskin. Reprint. (Englewood Cliffs, NJ: Prentice-Hall, 1966).
Christophers, Sir Richard S., *Aèdes Aegypti (L.) The Yellow Fever Mosquito. Its Life History, Bionomics and Structure* (Cambridge: Cambridge University Press, 1960).
Clément, Jean-Marc, "Histoire médicale de Saint-Louis," *Bulletin d'Epidemiologie*, no.9 (September 1985), pp. 210–212; no. 10 (October 1985), pp. 241–243; no. 11 (November 1985), pp. 274–276; no. 12 (December 1985), pp. 301–304.
Cohen, William B., "Malaria and French Imperialism," *Journal of African History*, 24 (1983), pp. 23–36.
Coleman, William, *Death is a Social Disease. Public Health and Political Economy in Early Industrial France* (Madison: University of Wisconsin Press, 1982).
Collignon, R., "Social Psychiatry in French-speaking Africa," in Olayiwola A. Erinosho and Norman W. Bell (eds.), *Mental Health in Africa* (Ibadan: Ibadan University Press, 1982).
Collignon, R., and Becker, C., *Santé et population en Sénégambie dès origines à 1960* (Dakar, 1989).
Collomb (Dr.), "Les maladies épidemiques en Afrique occidentale française en 1912. Fj.-Peste.-Variole," *Am. d'hyg. col*, t 17 (1914), pp. 940–954. extrait du Rapport annuel.
Colwell, Rita R., "Global Climate and Infectious Disease: The Cholera Paradigm," *Science* 274 (Dec. 20, 1996), 2026.
Commeleran (Dr.), "Sur un cas de délire systématise religieux. Rapport medico-legal," *Ann. d'hyg. col.*, t 17 (1914), pp. 130–140.
Condran, Gretcher A., and Eileen Crimmins-Gardener, "Public Health Measures and Mortality in U.S. Cities in the Late Nineteenth Century," *Human Ecology*, vol. 6, no. l, 1978, pp. 27–54.

Connolly, William E., *Identity/Difference: Democratic Negotiations of Political Paradox* (Ithaca, NY: Cornell University Press, 1991).

Cooper, Donald B., and Kenneth F. Kipple, 'Yellow Fever', in Kenneth F. Kipple (ed.), *The Cambridge World History of Human Diseases* (New York, 1993).

Cooper, F., *The Struggle for the City* (Beverly Hills CA: Sage Publications, 1983).

Coquery-Vidrovitch, C., "The Process of Urbanization in Africa," *African Studies Review*, vol. 34, no. 1 (April 1991), pp. 1–98.

Coquery-Vidrovitch, C., Histoire des villes d'Afrique noire : des origines à la colonisation (Paris: Albin Michel, 1993).

Corbin, Alain, *Les filles de noce. Misère sexuelle et prostitution aux 19e et 20e Siècles* (Paris: Aubier Montaigne, 1978).

Corbin, Alain, *Le miasme et la jonquille: l'odorat et l'imaginaire social, XVIIIe–XIXe siècles* (Paris: Aubier Montaigne, 1982).

Corre, A. (Dr.), "La matière médicale des noirs au Sénégal," *Moniteur du Sénégal et Dépendances*, nos. 1060, 6 June 1876, pp. 91–2; 1061, 13 June 1876, pp. 95–6; and 1062, 20 June 1876, pp. 99–100.

Courreges, Georges, and Fadel Dia, *Saint-Louis du Sénégal* (Clermont-Ferrand: Editions Soprep, 1982).

Coursier, A., *Faidherbe, 1818–1889: du Sénégal à l'armée du nord* (Paris: Tallandier, 1989).

Curson, Peter, and McCracken, Kevin, *Plague in Sydney. The Anatomy of an Epidemic* (Kensington, Australia: New South Wales University Press, 1989).

Curtain, Philip D., *Disease and Empire : The Health of European Troops in the Conquest of Africa* (Cambridge, U.K. ; New York : Cambridge University Press, 1998).

Curtin, Philip D., *Death by Migration. Europe's Encounter with the Tropical World in the Nineteenth Century* (Cambridge: Cambridge University Press, 1989).

Curtin, Philip D., "Medical Knowledge and Urban Planning in Tropical Africa," *American Historical Review*, 90 (3) (1985), 594–613.

Curtin, P. "'The White Man's Grave': Image and Reality," *Journal of British Studies*, 1(1961), pp. 94–110.

d'Anfreville, L., "La lutte contre les moustiques a Saint Louis du Sénégal," *Bull. Ste Path. Ex.*, 5 (1912), pp. 637–640.

Dedet, J.P., Les Instituts Pasteurs d'outre-mer: cent vingt ans de microbiologie française dans le monde (Paris: Éditions L'Harmattan, 2000).

de Vries, Jan, *European Urbanization, 1500–1800* (Cambridge: Harvard University Press, 1984).

Delacroix, Christian, Francois Dosse, and Patrick Garcia, *Les Courants historiques en France, XIXe–XXe siècle*. Éd. revue (Paris: Gallimard, 2005).

Delaporte, Francois, *The History of Yellow Fever: An Essay on the Birth of Tropical Medicine* (Cambridge, MA: MIT Press, 1991).

Delrieu, "Organisation du service de santé en Afrique occidentale française," *Ann. d'hyg. col.*, t 17 (1914), pp. 349–360. Extrait du rapport médical annuel de 1911.Denmark, R., and Thomas, K. P., "The Brenner-Wallerstein Debate," *International Studies Quarterly* 32 (1988): 47–65.

Deroure, F. "La vie quotidienne à Saint-Louis par ses archives, 1779–1809," *Bulletin de l'I.F.A.N.* B, T. 26.3-4 (1964).

Diop, A. S. "La foundation de Saint-Louis," *Bulletin de l'I.F.A.N.* T. 37.B.2 (1975), 1–50.

Diouf, M. "The French Colonial Policy of Assimilation and the Civility of the Originaires of the Four Communes (Senegal): A Nineteenth Century Globalization Project," *Development and Change* 29.4 (1998), 671–696.
Diouf, Mamadou, "Traitants ou négociants? Les commerçants Saint-Louisiens (2e moitié du XIXe s.–début XXe s.): Hamet Gora Diop (1846–1910). Etude de cas." Paper presented at the Colloque sur les Grands Commerçants Africains de l'Afrique Occidentale, Dakar, 1–4 May, 1990.
Dosse, François, *L'Histoire en miettes: Des "Annales" à la "nouvelle histoire"* (Paris: Éditions La Découverte, 1987).
Douglas, Mary. *Purity and Danger: An Analysis of the Concepts of Pollution and Taboo* (New York: Routledge, 2002).
Du Mazet, A., 'Note sur l'assainissement du Sénégal,' *Moniteur du Sénégal*, 21 February 1882, 31–33.
du Toit, Brian M., and Abdalla, Ismail H. (eds.), *African Healing Strategies* (New York: Trado-Medical Books, 1985).
Duval, M. J., *Les Colonies et la politique coloniale de la France* (Paris: Arthus Bertrand, 1864).
Echenberg, Myron. *Black Death, White Medicine: Bubonic Plague and the Politics of Public Health in Colonial Senegal, 1914–1945* (Oxford: James Curry, 2002).
Echenberg, M., *Colonial Conscripts. The Tirailleurs Sénégalais in French West Africa, 1857–1960* (Portsmouth, NH: Heinemann, 1991).
Eldredge, Elizabeth A., "Drought, Famine and Disease in Nineteenth-Century Lesotho," *African Economic History*, no. 16 (1987), pp. 61–93.
Ellis, Jack D., *The Physician-Legislators of France. Medicine and Politics in the Early Third Republic, 1870–1914* (New York: Cambridge U. Press, 1991).
Evans, R. J., *Death in Hamburg : Society and Politics in the Cholera Years, 1830–1910* (Oxford: Clarendon Press, 1987).
Faidherbe, L., *Le Senegal. La France dans l'Afrique Occidentale* (Paris: Librairie Hachette, 1889).
Farmer, Paul, *Infections and Inequalities: the Modern Plagues* (Berkeley: U. of California Press, 1999).
Fassin, D., *Pouvoir et maladie en Afrique. Anthropologie sociale dans la banlieue de Dakar* (Paris: PUF., 1992).
Faure, Claude, "Notice sur les archives du Sénégal," *Revue de l'Histoire des Colonies Françaises*, 2, juin 1914, pp. 363–374.
Feierman, S., "Struggles for Control: The Social Roots of Health and Healing in Modern Africa," *African Studies Review*, vol. 28 (1985), pp. 73–147.
Feierman, S., and Janzen, J. M. (eds.), *The Social Basis of Health and Healing in Africa* (Berkeley: University of California Press, 1992).
Fermer, F., Henderson, D. A., et al., *Smallpox and Its Eradication* (Geneva: WHO, 1988).
Flew, A. (ed.), *A Dictionary of Philosophy*. Rev. 2nd ed. (New York: St Martin's Press, 1979).
Ford, John, *The Role of the Trypanosomiases in African Ecology. A Study of the Tsetse Fly Problem* (Athens: Ohio University Press, 1971).
Foucault, Michel, *L'archéologie du savoir* (Paris: Gallimard, 1969).
Foucault, Michel, *Discipline and Punish* (Harmondsworth, 1979).
Foucault, Michel, *Folie et déraison. Histoire de la folie à l'âge classique* (Paris: Plon; 1961); réed. modifiée: *Histoire de la folie à l'âge classique* (Paris: Gallimard, 1972).

Foucault, Michel, *The History of Sexuality*. Translated from the French by Robert Hurley (New York: Penguin Books, 1976).

Foucault, Michel, *Il faut défendre la société, cours au Collège de France, 1974–1975* (Paris: Gallimard-Seuil-EHESS, 1997).

Foucault, Michel, *L'herméneutique du sujet, cours au Collège de France, 1981–1982* (Paris: Gallimard-Seuil-EHESS, 2001).

Foucault, Michel, *Histoire de la sexualité*, tome I: *La volonté de savoir* (Paris: Gallimard, 1976).

Foucault, Michel, *Histoire de la sexualité*, tome II: *L'Usage des plaisirs*, and tome III: *Le Souci de soi* (Paris: Gallimard, 1984).

Foucault, Michel, *Naissance de la biopolitique: Cours au Collège de France (1978–1979*, édition établie sous la direction de François Edwald et Alessandro Fontana (Paris: Seuil, 2004).

Foucault, Michel, *Naissance de la clinique. Une archéologie du regard médical* (Paris: P.U.F., 1963; rééd. 1972).

Foucault, Michel, *Surveiller et punir. Naissance de la prison* (Paris: Gallimard, 1975).

Gallagher, Nancy E., *Medicine and Power in Tunisia, 1780–1900* (Cambridge: Cambridge U. Press, 1983).

Gallagher, Nancy E., *Egypt's Other Wars. Epidemics and the Politics of Public Health* (Syracuse, NY: Syracuse U. Press, 1990).

Gallay, H. (Dr.), *Trois années d'assistance médicale aux indigènes et de lutte contre la variole. 1905. 1906. 1907* (Paris: Emile Larose, 1909).

Goldstein, Jan, *Console and Classify. The French Psychiatrie Profession in thé Nineteenth Century* (Cambridge: CUP, 1987).

Gouvernement Général de l'AOF. Colonie du Sénégal. *Principaux règlements concernant l'hygiène et la salubrité publiques* (SL: Imprimerie du Gvt, 1912).

Hamlin, Christopher, *Public Health and Social Justice in the Age of Chadwick, Britain, 1800–1854* (New York: Cambridge U. Press, 1998).

Hammond-Tooke, David, *Rituals and Medicines: Indigenous Healing in South Africa* (Johannesburg: Ad. Donker, 1989).

Harsin, Jill, *Policing Prostitution in Nineteenth-Century Paris* (Princeton, NJ: Princeton University Press, 1985).

Hartwig, G. W., and Patterson, K. D. (eds.), *Disease in African History. An Introductory Survey and Case Studies* (Durham, NC: Duke University Press, 1978).

Headrick, Rita, *Colonialism, Health and Illness in French Equatorial Africa, 1885–1935* (Atlanta: African Studies Association Press, 1994).

Headrick, Daniel R., *The Tools of Empire. Technology and European Imperialism in the Nineteenth Century* (New York: Oxford University Press, 1981).

Hehnan, Cecil G., *Culture, Health and Illness. An Introduction for Health Referrals.* 2nd ed. (Boston: Butterworks Co., 1990).

Henige, D., *Oral Historiography* (London: Longman, 1982).

Hohenberg, Paul H., and Lees, Lynn Hollen, *The Making of Urban Europe, 1000–1950* (Cambridge, MA: Harvard University Press, 1985)

Hopkins, Terence K., and Wallerstein, I. (eds.), *World-Systems Analysis: Theory and Methodology* (Beverly Hills, CA: Sage, 1982).

Houillon (Dr.), "Variole et vaccine en Afrique occidentale française pour l'année 1903," *Ann. d'Hyg. Col.*, T 8 (1905), pp. 546–568.

Humphries, S., *The Handbook of Oral History. Recording Life Stories* (London: Inter-Action Imprint, 1984).
Hunt, Nancy Rose, *A Colonial Lexicon of Birth Ritual, Medicalization, and Mobility in the Congo* (Durham, NC: Duke University Press, 1999).
Idowu, H. O., "The Establishment of Elective Institutions in Senegal, 1869–1880," *Journal of African History* 9 (1968), 261–77.
James, Tracy, D. (ed.), *The Rise of Merchant Empires. Long-Distance Trade in the Early Modern World, 1350–1750* (Cambridge: Cambridge University Press, 1990).
Janzen, John M., *The Quest for Therapy. Medical Pluralism in Lower Zaire* (Berkeley: UCP, 1978).
Janzen, John M., "The Need for a Taxonomy of Health in the Study of African Therapeutics," *Social Science and Medicine*, 15B, 3 (1981), p. 190.
Jennings, Eric T., *Curing the Colonizers: Hydrotherapy, Climatology, and French Colonial SPAS* (Durham and London: Duke University Press, 2006).
Johnson, G. W., *The Emergence of Black Politics in Senegal, the Struggle for Power in the Four Communes, 1900–1920* (Stanford, CA: Stanford U. Press, 1971).
Johnson, G. W., *Naissance du Sénégal contemporain. Aux origines de la vie politique moderne, 1900–1920* (Paris: Karthala, 1991).
Kasaba, R. (ed.), *Cities in the World-System* (New York: Greenwood Press, 1991).
Kermorgant, A., "Relation d'une enquête relative a la maladie du sommeil dans le gouvernement general de l'Afrique occidentale française," *Ann. d'Hyg. Col*, t 7 (1904), pp. 274–284.
Kermorgant, A. (Dr.), "Epidémie de fievre jaune du Sénégal, du 16 avril au 28 fev 1901," *Ann. d'Hgy. Col*, t 4 (juillet–aout–sept. 1901), pp. 325–436.
Kermorgant, A. (Dr.), "Aperçu sur les maladies vénériennes dans les colonies françaises," *Ann. d'hyg. col*, t 6 (1903), pp. 428–460.
Kermorgant, A. (Dr.), and Reynaud, G., "Précautions hygieniques a prendre pour les expéditions et les explorations aux pays chauds," *Ann. d'hgy. col*, t 3 (jt–aout–sept 1900), pp. 305–414.
Khun, Thomas S., *The Structure of Scientific Revolutions*. 2nd ed. (Chicago: University of Chicago Press, 1970).
King, A., *Global Cities. Post-Imperialism and the Internationalization of London* (New York: Routledge 1990).
King, Anthony, D., *Urbanization, Colonialism, and the World-Economy. Cultural and Spatial Foundations of the World Urban System* (New York: Routledge, 1900).
Kleinman, Arthur, *The Social Origins of Distress and Disease* (New Haven, CT: Yale University Press, 1986).
Knaut, Andrew L., "Yellow Fever and the Late Colonial Public Health Response in the Port of Veracruz," *Hispanic American Historical Review* 77, 4 (Nov. 1997), 619–644.
Knight, Franklin W., and Peggy K. Liss (eds.), *Atlantic Port-Cities: Economy, Culture, and Society in the Atlantic World, 1650–1850* (Knoxville: University of Tennessee Press, 1991).
Latour, Bruno, *The Pasteurization of France*. Trans. by Alan Sheridan and John Law (Cambridge, MA: Harvard U. Press, 1988).
Lawrence, A. W., *Trade Castles and Forts of West Africa* (London: Jonathan Cape, 1963), 78.
Lederberg, J., R. E. Shope, and S. C. Oaks Jr. (eds.), *Emerging Infections. Microbial Threats to Health in the United States* (Washington, DC: Institute of Medicine, National Academy Press, 1992).

Leonard, J., "Les Officiers de santé de la Marine française de 1814 a 1835," (Thèse 3e cycle, 1967), p. 335.

Leavitt, Judith W., *The Healthiest City. Milwaukee and the Politics of Health Reform* (Princeton, NJ: Princeton U. Press, 1982).

Loti, Pierre, *Correspondence inédite, 1865–1904* (Paris: Calman-Levy, 1929).

Lowy, Ilana, "Yellow Fever in Rio de Janeiro and the Pasteur Institut Mission (1901–1905): The Transfer of Science to the Periphery," *Medical History* 34 (1990), 144–163.

Lyons, Maryinez, *The Colonial Disease. A Social History of Sleeping Sickness in Northern Zaire, 1900–1940* (New York: Cambridge U. Press, 1992).

M'Bokolo, E., "Peste et société urbaine à Dakar: l'épidémie de 1914," *C.E.A.*, XXII, 85–86, 1982, pp. 13–46.

MacLeod, Roy, and Lewis, Milton (eds.), *Disease, Medicine, and Empire. Perspectives on Western Medicine and the Experience of European Expansion* (New York: Routledge, 1988).

Manchuelle, François, "Métis et colons: la famille Devès et l' emergence politique des Africains au Sénégal, 1881–1897," *Cahiers d'Études Africaines* 96.XXIV-4 (1984), 477–504.

Mandeleau, Tita, *Signare Anna ou le voyage aux escales* (Dakar: NEAS, 1991).

Marc Michel, "Le Corps de Santé des troupes coloniales," pp. 185–213: Ecole du Pharo à Marseille (Institut de Médecine Tropicale), 1905–).

Marchoux, B., *Traité de Pathologie Exotique, Chimique et Thérapeutique* (sous la direction de C. Grail, A. Clarac) (Paris: Baillere & Fils, 1910).

Marchoux, E., and Bourret, G., "Recherches sur la transmission de la lèpre," *Ann. d'Hyg. Col.*, t 13 (1910), pp. 169–172. (See *Ann. de l'Institut Pasteur*, t. XXIII, jt 1909).

Marchoux (Dr.), "Fonctionnement du laboratoire de microbiologie de Saint-Louis (Senegal) et Note sur la dysenterie des pays chauds," *Ann. d'Hyg. Col.*, 3 (1900), pp. 119–131.

Marchoux (Dr.), "Au sujet de la transmission du paludisme par les moustiques," *Ann. d'Hyg. Col.*, t 2 (1899), pp. 22–25.

Marchoux, E., Pl Simond, A. Salimbeni, "Etudes sur la fièvre jaune," premier mémoire, *Annales de l'Institut Pasteur* 17 (Nov. 1903, 665–728); Idem., 2me mémoire, *Annales de l'Institut Pasteur* 25 (Janv. 1906), 16–40.

Martin, Phyllis. *Leisure and Society in Colonial Brazaville* (New York: Cambridge University Press, 1995).

Massiou (Dr.), "La vaccine à Saint-Louis (Senegal)," *Ann. d'Hyg. Col.*, t 7 (1904), pp. 17–19.

Mathis, Constant, *L'Oeuvre des Pastoriens en Afrique Noire* (Paris: PUF, 1946).

Mauny, R., "Pierre Loti au Sénégal (1873–1874)," *Notes Africaines*, no. 74, April 1957, pp. 55–61.

Maurice, *Notions d'hygiène et de médecine à l'usage des colonies* (Paris, 1920).

Mbiti, John S., *African Religions and Philosophy* (Oxford: Heinemann International, 1989).

McKelvey Jr., John J., *Man Against Tsetse. Struggle for Africa* (London: Cornell University Press, 1973).

McNeill, William, *Plagues and Peoples* (Garden City, NY: Anchor Press, 1976).

Merveilleux (Dr.), "Notes sur la situation sanitaire du Senegal pendant l'année 1909 (Morbidité et mortalité, mouvement de la population, etc.), *Ann. d'Hyg. Col.*, t XXin, jt 1909, pp. 676–704 (extrait du rapport annuel, 1909).

Merveilleux (Dr.), "Notes démographiques et protection de l'enfance à Saint-Louis," *Ann. d'Hyg. Col.*, t 9 (1906), pp. 132–139.

Miller, J. C. (éd.), *The African Past Speaks: Essays on Oral Tradition and History* (Folkestone, England: Dawson, 1980).

Ministère de l'Urbanisme et du Logement, *Urbanisme et habitat en Afrique noire francophone avant 1960* (Paris: Agence Française pour l'Aménagement et le Développement a l'Etranger, April 1984).

Moal (Dr.), "Etude sur les moustiques en Afrique occidentale française (rôle pathogenique.–prophylaxie)," *Ann. d'Hyg. Col*, t 9 (1906), pp. 181–219.

Moulin, Anne Marie, "Patriarchal Science: The Network of Overseas Pasteur Institutes," in Patrick Petitjean, Catherine Jami, and Anne Marie Moulin (eds.), *Science and Empires: Historical Studies about Scientific Development and European Expansion* (Dordrecht; Boston: Kluwer Academic Publishers, 1992), 307.

Navarre, Just P. (Dr), *Manuel d'hygiène coloniale. Guide de l'Européen dans les pays chauds* (Paris: Octave Doin, 1895).

Needell, Jeffrey D., "The Revolta Contra Vacina of 1904: The Revolt against 'Modernization' in Belle-Époque Rio de Janeiro," *Hispanic American Historical Review*, vol. 67, no. 2 (May 1987), 233–269.

Ngalamulume, Kalala, « City Growth, Health Problems, and the Colonial Government Intervention in Saint-Louis-du-Senegal, from Mid-Nineteenth to the Early Twentieth C. (Ph.D. dissertation, Michigan State University, 1996).

Ngalamulume, Kalala, « Keeping the City Totally Clean : Yellow Fever and the Politics of Prevention in Saint-Louis-du-Senegal , 1850-1914, » *Journal of African History* 45.2 (2004), 183-202.

Ngalamulume, Kalala, "Classify and Sequestrate: The Regulation of Madness in Saint-Louis-Du-Senegal, 1890–1914," in Kalala Ngalamulume and Paula Viterbo (eds,), *Health and Medicine in Africa* (Berlin: LIT Verlag, 2010).

Niane, D. T., and Suret-Canale, J., *Histoire de l'Afrique occidentale* (Paris: Présence Africaine, 1961).

Niang Fatou Siga, *Reflets de modes et traditions saint louisiennes* (Dakar: Ed. Khoudia, 1990).

Nicolas, P., *Bioclimatologie humaine de Saint-Louis du Sénégal (Essai de méthodologie bioclimatologique)* (Dakar: IFAN, 1959).

Nicolas, B., *Manuel d'hygiène coloniale* (1894).

Nuttall, Sarah, and Achille Mbembe (eds.), *Johannesburg: The Elusive Metropolis* (Durham, NC: Duke University Press, 2008).

O'Connor, A., *The African City* (London: Hutchinson U. Library for Africa, 1983).

Oldstone, M., *Viruses, Plagues, and History* (New York: Oxford U. Press, 1998).

Packard, R., *White Plague, Black Labor. Tuberculosis and the Political Economy of Health and Disease in South Africa* (Berkeley: UCP, 1989).

Pasquier, P., "Un aspect de l'histoire des villes du Sénégal: les problèmes de ravitaillement au XIXe siècle," *Cahiers du C.R.A.*, no. 5 (1987), p. 192.

Pasquier, Roger. "Villes du Sénégal au XIXe siècle," *Revue d'Histoire d'Outre-Mer* T. XLVII (1960), 387–426.

Patterson, K. D., "Disease and Medicine in African History: A Bibliographical Essay," *History in Africa* 1(1974), pp.141-148.

Pelletier, J. (Dr.), "Cas d'elephantiasis du scrotum observés au Sénégal," *Bull. Ste Path. Ex.*, 5 (1912), pp. 625–627.

Perotin-Dumon, A., "Cabotage, Contraband, and Corsairs: The Port Cities of Guadeloupe and Their Inhabitants, 1650–1800," in Knight, F. W. and Liss, Peggy K., (eds.), *Atlantic Port Cities. Economy, Culture, and Society in the Atlantic World, 1650–1850* (Knoxville: U. of Tennessee Press, 1991).

Peyrot, (Dr.), "Us, coutumes, médecine des Bambaras," *Ann. d'Hyg. Col.*, t 8 (1905), pp. 456–473.

Pluchon, P. (sous la dir de), *Histoire des médecins et pharmaciens de Marine et des colonies* (Toulouse: Privât, 1985).

Polychroniou, P., *Marxist Perspectives on Imperialism. A Theoretical Analysis* (New York: Praeger, 1991).

Pressman, Jack D., "Concepts of Mental Illness in the West," in Kenneth F. Kipple (ed.), *The Cambridge World History of Human Disease* (New York: Cambridge University Press, 1993).

Preston, Samuel H., and Etienne Van De Walle, "Urban French Mortality in the Nineteenth Century," *Population Studies*, 32.2 (July 1978), pp.275–297.

Prochaska, D., *Making Algeria French. Colonialism in Bone, 1870–1920* (New York: Cambridge University Press, 1990).

Provost, *Assistance Indigène* (Paris, 1909).

Pulvenis, Claude, "Une epidémie de fièvre jaune à Saint-Louis du Sénégal (1881)," *Bulletin de l'Institut Fondamental de l'Afrique Noire*, B, XXX (1968): 1353–1373.

Quella-Villeger, Allain, *Pierre Loti l'incompris* (Paris: Presses de la Renaissance, 1986).

Ransford, Oliver, *'Bid the Sickness Cease.' Disease in the History of Black Africa* (London: Oliver Ransford, 1983).

Reed, Walter, "Recent Researches Concerning the Etiology, Propagation, and Prevention of Yellow Fever, by the United States Army Commission," *Journal of Hygiene* 2.2 (April 1, 1902), 104–107, 111–15, 118–19.

Reeves, Peter, "Studying the Asian Port City," in Frank Broeze (ed.), *Brides of the Sea. Port Cities of Asia From the 16th–20th Centuries* (Honolulu: University of Hawaii Press, 1989).

Reid, Donald. *Paris Sewers and Sewermen: Realities and Representations* (Cambridge, MA: Harvard University Press, 1991).

Renneville, Marc, "Politiques de l'hygiène de l' AFAS (1872–1914), in Patrice Bourdelais (dir.), *Les Hygiénistes: enjeux, modèles et pratiques* (Paris: Éditions Belin), 2001.

Reynaud, G., *Hygiène des colons* (Paris, 1903).

Reynaud, G., *Hygiène des etablissements coloniaux* (Paris, 1903).

Robinson, David, *Paths of Accommodation: Muslim Societies and French Colonial Authorities in Senegal and Mauritania, 1880–1920* (Athens: Ohio University Press, 2000).

Robinson, David, et Triaud, Jean-Louis (éds.), Le Temps des marabouts : itinéraires et stratégies islamiques en Afrique occidentale française v. 1880-1960 (Paris : Éditions Karthala, 1997).

Rose, Jerry D., *Outbreaks. The Sociology of Collective Behavior* (New York: Free Press, 1982).

Rosenberg, Charles F., *The Cholera Years: The United States in 1832, 1849, and 1866* (Chicago: University of Chicago Press, 1962).

Rosenberg, Charles. "What is an Epidemic? AIDS in Historical Perspective," *Daedalus*, 118 (1989), 1–17.

Rosenberg, Charles., and Golden, Janet, Framing Disease: Studies in Cultural History (New Brunswick, N.J.: Rutgers University Press, 1992).

Ross, R., and Telkamp, J., *Colonial Cities. Essays on Urbanism in a Colonial Context* (Dordrecht, Netherlands; Boston: M. Nijhoff, 1985).

Rousseau, R. "Le site et les origines de Saint-Louis," *La Géographie*, vol. 44, nos. 2–3–4–5 (July, Sept., Nov., Dec. 1925).
Russell, Bertrand, *A History of Western Philosophy* (New York: Simon and Schuster, 1972).
Saint-Vincent, Y.-J., *Le Sénégal sous le Second Empire. Naissance d'un empire (1850–1871)* (Paris: Karthala, 1985).
Sanarelli, J. (Dr.), "La Fièvre jJaune," *Monographie*, no 8 (15 avril 1898), pp. 280–314.
Sanarelli, J. (Dr.), "La fièvre jaune," *L 'Oeuvre médico-chirurgicale*, no.8 (April 1898), pp. 3–4.
Saulnier, E., *La Compagnie de Galam au Sénégal* (Paris: Emile Larose, 1921).
Sawchuk, Lawrence A., "Deconstructing an Epidemic: Cholera in Gibraltar," in D. Ann Herring and Alan C. Swedlund (eds.), *Plagues and Epidemics: Infected Spaces Past and Present* (New York: Berg, 2010).
Scale, Clive, and Pattison, Stephen (eds.), *Medical Knowledge: Doubt and Certainty* (Philadelphia: Open University Press, 1994).
Simon, David, *Cities, Capital and Development. African Cities in the World Economy* (London: Belhaven Press, 1992).
Singer, Merill, "The Coming of Age of Critical Anthropology," *Social Science and Medicine*, 28.11 (1989), p.
Sinou, A., *Comptoirs et villes coloniales du Sénégal: Saint-Louis, Corée, Dakar* (Paris: Ed. Karthala, 1993).
Smelser, Neil J., *Theory of Collective Behavior* (New York: Free Press of Glencoe, 1963).
Smith, Alan K., *Creating A World Economy. Merchant Capital, Colonialism, and World Trade, 1400–1825* (Boulder, CO: Westview Press, 1991).
Snowden, F. M., *Naples in the Time of Cholera, 1884–1911* (New York: Cambridge University Press, 1995).
Sole, Jacques, *L'âge d'or de là prostitution: De 1870 à nos jours* (Paris: Pion, 1993).
Soleillet, P., *Voyage à Segou, 1878–1879* (Paris: Challamelaine, 1887).
Strath, Bo (ed.), *Language and the Construction of Class Identities: The Struggle for Discursive Power in Social Organization: Scandinavia and Germany after 1800* (Gothenberg: Gothenberg U., 1990).
Strong, Phillip, 'Epidemic Psychology: A Model', *Sociology of Health and Illness* 12.3 (1990), 249–59.
Stub, Holger R. (ed.), *Status Communities in Modern Society. Alternatives to Class Analysis* (Hinsdale, IL.: Dryden Press, 1972).
Swanson, Maynard, "The Sanitation Syndrome: Bubonic Plague and Urban Native Policy in the Cape Colony, 1900–1909," *Journal of African History*, XVIII, 3 (1977), pp. 387–410.
Swanson, Maynard W., "'The Asiatic Menace': Creating Segregation in Durban, 1870–1900," *International Journal of African Historical Studies*, 16, 3 (1983), pp. 401–421.
Tardieu, À., *L'Univers. Histoire et description de tous les peuples. Senegambie, Guinée, Nubie, Âbyssinie* (Paris: Firmin Didot Frères, 1847).
Taussig, M., "Reification and the Consciousness of the Patient," *Social Science and Medicine*, 14b (1980), pp. 3–14.
Taussig, M., *Shamanism, Colonialism and the Wild Man* (Chicago: University of Chicago Press, 1987).

Telkamp, G. J., *Urban History and European Expansion. A Review of Recent Literature Concerning Colonial Cities and a Preliminary Bibliography* (Leiden Centre for the History of European Expansion, 1978).

Tevera, D. S., et al. (ed.), *Harare. The Growth and Problems of the City* (Harare: University of Zimbabwe Press, 1993).

Thiroux, A., "Le N'diank, cholera du Sénégal; Son agent pathogène," *Bull. Ste Path. Exot.*, 5 (1912), pp. 753–762.

Thiroux, A. (Dr.), "Les villages de segrégation et de traitement de la maladie du sommeil. Fonctionnement d'un de ces villages à Saint-Louis-du-Sénégal," *Ann. d'Hyg. Col.*, t 12, 1909, pp. 448–459.

Thiroux, Dr. A., and Anfreville, Dr. L. D', *Le paludisme au Sénégal pendant les années 1905–1906* (Paris: Libr J.-B. Bailliere et Fils, 1908).

Thiroux, A., and L. d'Anfreville, *Le paludisme au Sénégal pendant les années 1905–1906* (Paris: Libr. J.B. Bailliere et Fils, 1908).

Thomas, Lynn M., *Politics of the Womb: Women, Reproduction, and the State in Kenya* (Berkeley: University of California Press, 2003)

Thoulon, rapport sur l'epidemie d'influenza de 1918–1919, pp. 270–276.

Timberlake, M. (ed.), *Urbanization in thé World Economy* (Orlando, FL: Academic Press, 1985).

Löwy, I., "Yellow Fever in Rio de Janeiro and the Pasteur Institute Mission (1901-1905): the Transfer of Science to the Periphery," Med Hist. 34.2 (April 1990), 144-163.

Troupes Coloniales. Revue Historique de l'Armée. Les troupes de marine, 1870–1970 (Paris: Ministère de la Défense Nationale, 1970)

Turshen, Meredith, *The Political Ecology of Disease in Tanzania* (New Brunswick, NJ: Rutgers University Press, 1984).

Twaddle, A. C., and R. N. Hessler, *A Sociology of Health*. 2nd ed. (New York: Macmillan, 1987).

Vansina, J., *Oral Tradition as History* (Madison: University of Wisconsin Press, 1985).

Vaughan, Megan, *Curing Their Ills. Colonial Power and African Illness* (Stanford: Stanford University Press, 1991), pp. 204–S.

Wallerstein, I., *The Modern World System I: Capitalist Agriculture and the Origin of the European World-Economy in the Sixteenth Century* (New York: Academic Press, 1974).

Wallerstein, I., *The Modern World-System II: Mercantilism and the Consolidation of the European World Economy, 1600–1750* (New York: Academic Press,'1980).

Wallerstein, I., *The Modern World-System III: The Second Era of Great Expansion of the Capitalist World-Economy, 1730–1860s* (New York: Academic Press, 1989).

Wallerstein, I. "The Itinerary of the World System Analysis, or How to Resist Becoming a Theory," in J. Berger and M. Zelditch Jr. (eds.), *New Directions in Sociological Theory* (Lanham, MD: Rowman & Littlefield Publishers, 2002), 359–376.

Webb Jr., J. L. A., "Malaria in Early Tropical Africa," in Paula Viterbo and Kalala Ngalamulume (eds.), *Health and Medicine in Africa: Multidisciplinary Perspectives* (Berlin: LIT Verlag, 2010; East Lansing: Michigan State University Press, 2010).

Werner, J.-F., *Marges, sexe et drogues à Dakar. Ethnographie urbaine* (Paris: Karthala, 1993).

Wohl, Anthony S., *Endangered Lives. Public Health in Victorian Britain* (Cambridge, MA: Harvard U. Press, 1983).

Wondji, C., "La fièvre jaune à Grand-Bassam (1899–1903), *Revue fr. d'Hist. d'O.M.*, T. LIX, no. 215 (1972), pp. 205–239.

Zeleza, Paul Tiyambe, and Cassandra Rachel Veney, *Leisure in Urban Africa* (Trenton, NJ: Africa World Press, 2003).
Zuccarelli, F., *La vie politique sénégalaise, 1789–1940* (Paris: C.H.E.À.M., 1988).
Zuccarelli, F., "Les maires de Saint-Louis et Gorée de 1816 à 1872," *Bulletin de l'I.F.A.N*, B, vol. 35, no. 3 (1973).
Zuccarelli, F., "Les traitants des comptoirs du Sénégal au XIXe siècle," *Entreprises et Entrepreneurs en Afrique* (Dec. 1981).

# Index

500–800 first plague pandemic, 186
1348–1700 second plague pandemic, 186
1817–1823 first cholera pandemic, 89
1829–1851 second cholera pandemic, 89
1852–1859 third cholera pandemic, 89
1863–1879 fourth cholera pandemic, 89. *See also* 1868 cholera epidemic
1867 yellow fever epidemic, 61
1868 cholera epidemic
   in fourth cholera pandemic, 89, 96–97
   measures taken, 98
   mortality statistics, 61, 63, 97–100
   overview of, 96–101
   panic, 98
   treatments, 98–99
1869 cholera epidemic, 102
1878 yellow fever epidemic
   Berenger-Féraud's germ theory, 57
   monument to victims (photo), 162
   overview of, 61–62
   panic response, 69
   preventive measures, 73–74
1880–81 yellow fever epidemic
   Berenger-Féraud's germ theory, 57–58

1880–81 yellow fever epidemic, *cont.*
   contagionist and localist theories for, 58
   overview of, 62–63, 134–135
   panic response, 69
   preventive measures, 73–74
1881–1896 fifth cholera pandemic
   1893 cholera epidemic in, 109–114
   overview of, 90
   in Saint-Louis, 92
   system, 105–109
1893 cholera epidemic
   aftermath, 114–115
   conflict over quarantine measures, 129
   overview of, 109–114
1894–1948 third plague pandemic, 186
1899–1923 sixth cholera pandemic, 90
1900 yellow fever epidemic
   blaming women for, 72
   conflicts of interests, 131–137
   fear and panic during, 67–68
   indigenous society believed immune to, 66–67
   monitored from Paris, 63–64
   official declaration of, 65

1900 yellow fever epidemic, *cont.*
   problems diagnosing, 63, 68
   protests against anti-yellow fever measures, 137–140
   repatriation of Europeans in, 65–67
   sanitary measures and, 64, 68
   scientific missions to Brazil. *see* Simond-Marchoux-Selimbeni Mission in Brazil (1901–1905)
   scientific missions to Senegal. *see* Grall-Marchoux-Jacquerez Sanitary Mission to Senegal (1901)
   temporary subsiding of, 64–65
1961–present seventh cholera pandemic, 90

# A

abbreviations, used in this book, xv
accommodationist strategy, plague policy
   biopolitical strategies vs. sanitarian, 190–191
   Blaise Diagne's intervention, 191–193
   Carrera's intervention, 191
   end of, August–November 1918, 200–204
   Governor Lévecque's dilemma, 195–196
   mortality, fear, protest, and ìradical solution,î 197–200
   outbreak in Saint-Louis, 187–188
   popular protest, 188–190
   radical measures of Colonial Hygiene Committee, 193–195
accusations
   as panic response to yellow fever, 69–70
   protests against quarantine measures, 128
   scapegoating victims, 209–210
acquired immunity
   defined, 52
   to yellow fever, 80
ACSE (Archives of the Congregation of Saint-Esprit), 10
*Aedes aegypti* mosquitoes
   dengue fever from, 32
   ecology of yellow fever, 51, 141–142
*Aedes Africanus* mosquitoes, 51
*Aedes Albopictus* mosquitoes, 32

AFAS (*Association Française pour l'Avancement des Sciences*), 141
Africa
   1868 cholera traveling from Northern, 100
   yellow fever origins in, 52
Agramonte, d'Aristide, 141
Americas, yellow fever origins in, 52
Angoulvant, Governor General, 197–200
*Annexe de Saint-Louis*, 37
*Annexe Notre-Dame-de-Lourdes*, Sor, 41
*Anopheles* mosquitoes, 32
antibiotics, bubonic plague treatment, 186
anti-plague serum, 186, 187
Archives of the Congregation of Saint-Esprit (ACSE), 10
*Assistance Médicale Indigène*, 10, 37
August 29, 1884 decree
   1899 yellow fever debate, 132–137
   compromise between contagionists/merchants, 129, 137
   fifth cholera pandemic and, 110

# B

Bacteriological Laboratory of Saint-Louis (1896), 140
bacteriological revolution
   bubonic plague practices despite, 209
   commercial interests strengthened by, 137
   stigmatization of urban poor despite, 12
bacteriology, science of
   clinging to older models of disease vs., 2, 50
   and control of yellow fever, 140–142
   sanitary measures for cholera informed by, 90
Badois contract, water system, 106–107
Ballay, Governor General
   1900 yellow fever epidemic, 144–146
   applying new knowledge, 175
   repatriation of Europeans in 1900 yellow fever, 67–68
   tension over quarantine with Combes, 138–139

Bamana (Bambara)
  conversion to Christianity, 24
  docile acceptance of anti-plague measures, 189
  living in center of city as domestic servants, 25
  living in Sor, 26
  proposal of ethnic-based village for, 194
  settling in Saint-Louis, 22–23
Bay of Bengal, as origin of cholera, 88
Bérenger-Féraud, Dr. L.-J.-B.
  on causes of *hivernage*-related diseases, 31
  on germ theory for transmission of yellow fever, 57
  on scientific basis for harmful climate of Senegal, 54–55
biopolitical strategies, 1917 plague outbreak, 190–191
"Black Man's Disease." *see* cholera epidemics
Blanc, J. Le, 107–108
Borius, Dr. A, 54–55
Bouetville, Sor
  creating village in 1852 at, 24
  draining swamp at, 171, 177
  enforcing local hygiene in slums of, 123
  fire regulations in, 27
  village for sleeping sickness patients at, 41
Bouley, Henri-Marie, 141
Bouna, Moctar, 196
Brazil
  bad reputation of Rio de Janeiro in, 17
  dispatching scientific mission to. *see* Simond-Marchoux-Selimbeni Mission in Brazil (1901–1905)
  sanitary measures for ships arriving from, 121
  yellow fever epidemics originating in, 52, 62, 148
bubonic plague. *see* plague and violence (1917–1920)
builders (*entrepreneurs de travaux publics*), draining swamps, 179–181
built environment
  colonial city of Saint-Louis, 24–27
  eliminating dangerous classes and noise from, 42–43

built environment, *cont.*
  reflecting historical evolution/occupational structure, 27–28
Buléon, Bishop, 65
*Bulletin de la Congrégation de Saint-Joseph de Cluny* (1885–1904), 10
burial
  1775 warnings about decomposing bodies, 53
  1918 anti-plague measures, 193
  avoiding infection during ritual of, 190
  biopolitical vs. sanitarian for plague control at, 191
  cemeteries as source of infectious miasmas, 74
  cholera transmission during rituals of, 92, 111
  dispute between Freemason and Catholics, 28
  obtaining mayorial authorizations for, 74
  yellow fever guidelines for speedy, 140, 175

## C

Carpot, Dr. Charles
  combating 1893 cholera epidemic, 110–111
  documenting *Hospice Civil*, 48
  return to Senegal after medical school, 80
  uncertain cause of yellow fever, 165
Carpot, François, 188, 196
Carpot, Pierre, 69–70
Carpot, Theodore, 173
Carpot family, 20–22, 29
Carrera, Administrator of Colonies, 191, 202–203
Carroll, James, 141
Catholic nuns
  distribution in medical facilities, 40
  nursing care of the Sisters, 36
  opening Sainte Anne dispensary in Ndar-Toute, 41
Catholics
  church response to cholera epidemic, 101
  fighting anti-clericalism of Freemasons, 28
  response to yellow fever epidemics, 69–72

cemeteries
   1867 yellow fever epidemic and changes in, 61
   1881 yellow fever epidemic and, 58
   danger as source of miasmas, 56
   yellow fever preventive measures, 74
Central Place System, urban historical model, 6
*Cercle de la Concorde*, 28
*Cercle des Bons Amis*, 29
*Cercle des Habitants Notables du Pays*, 29
*Cercle du Progrès*, 29
CFAO (*Compagnie Française de l'Afrique Occidentale*), 142–144
Chaudié, Governor General
   1899 yellow fever outbreak and, 131–137
   criticisms of in 1900 epidemic, 146–147, 164–165
   protests against abuse of, 139–140
Chief Medical Officer. *see Médecin en Chef*
children's immunity
   to malaria, 31
   to plague, 186
   to yellow fever, 165, 174
Chimère, Pierre, 29
cholera epidemics
   1868. *see* 1868 cholera epidemic
   1869, 102
   1893, 109–115, 129
   conclusion, 115–116
   early efforts to build waterworks, 96
   ecology of, 90–93
   origin and spread of, 88–90
   overview of, 88
   political impact of, 103–105
   from polluted water, 33
   recorded pandemics, 89–90
   sewerage problems, 93–94
   water supply problems, 94–96
   water supply system, 105–109
CHPS (Council of Hygiene and Public Salubrity), 61
Civilian Hospital
   creating for poor, 37
   distribution of Catholic nuns in, 40

Civilian Hospital, *cont.*
   government cisterns meeting needs of, 77
   mentally ill patients temporarily kept in, 42
   not containing all patients during epidemics, 77
   unequal access to health care and, 35
   yellow fever mortality statistics recorded in, 60
classes, nursing profession, 40
climate. *See also hivernage*-related diseases
   ecology of yellow fever, 51–52
   etiology of yellow fever, 59–60
   intestinal system diseases and, 32–33
   scientific basis for harmful Senegal, 54–55
clothing
   compulsory disinfection of contaminated, 129, 133, 140
   contagionist theory of yellow fever from, 56, 63–64, 133, 168
   destruction of mosquitoes to prevent yellow fever vs., 142
   disinfection of contaminated, 65, 75, 77
   spread of cholera via, 90–91, 92, 95, 111
   spread of smallpox via, 33
Colonial Hospital
   categories of patients in, 37
   distribution of Catholic nuns in, 40
   military hygiene and, 167, 169
   overcrowding and hospital infection in, 35
   transforming Military Hospital into, 37
Colonial Hygiene Committee
   Angoulvant rejects strategy of, 197
   Lévecque rejects proposed measures, 195–196
   radical plague measures of, 193–195
Combes, General, 67, 138–139
commercial considerations
   economic consequences of yellow fever, 143
   justifying scientific mission to Senegal, 165
*Commissaire de la République*, 191–193
*Compagnie Française de l'Afrique Occidentale* (CFAO), 142–144
conflict of interests
   bacteriology and control of yellow fever, 140–142
   conclusion, 149–150

conflict of interests, *cont.*
  contagionists and quarantine measures, 124–126
  crisis of confidence, 131–137
  injunctions to prevent disease outbreaks, 120–121
  localists and urban hygiene, 121–124
  protests against 1900 yellow fever measures, 137–140
  protests against automatic quarantine, 126–131
  toward new paradigm, 142–149
*Conseil de Défense* (Defense Council), 50, 79
*Conseil Privé* (Private Council)
  declaring yellow fever epidemic in 1900, 65
  membership of Médecin en Chef into, 50, 79–80
contagionists
  complementary with localists, 58–59
  conception of 1868 cholera epidemic, 100
  conception of yellow fever spread, 56–57, 73–76, 141–142
  conflict of interests with, 136–137
  opposition of anti-contagionists with, 132–137
  quarantine measures and, 124–126
  skepticism about disinfection credo of, 133
Contaut, Dr., 194–197
Corbin, Alain, 53
Creoles
  administration vs. municipality authority of, 173
  children lacking immunity to yellow fever, 165
  role during formation of French West Africa, 7

# D

d'Azyr, Vicq, 53
death penalty, bureaucrat failure in combating epidemics, 77, 130
Decrais, Albert, 131–134, 145–149
Defense Council (*Conseil de Défense*), 50, 79
demographics, early Saint-Louis, 19

dengue fever, 32
Devès, Mayor Gaspard, 70
Devès, Mayor Justin, 181
Diagne, Blaise
  1918 plague policy intervention, 191–192
  elected deputy in 1914, 29
  failure to convince leaders of Guet-Ndar to comply, 200
diarrhea
  caused by polluted water, 33
  in cholera. see cholera epidemics
  *hivernage*-related, 17, 31, 55, 96
  Saint-Louis water containing pathogens for, 114
  from yellow fever in 1879, 69
Diaw, Coumba, 189–200
Diop, Hamet Gora, 23
disinfection (fumigation)
  1893 cholera epidemic, 110–111
  1900 findings in yellow fever opposing, 142
  1902 decree for measures against epidemic diseases, 175
  1904–05 sanitation ordinances of Governor Guy, 177
  1917 plague outbreak in Saint-Louis, 187–188
  Colonial Hygiene Committee's radical plague measures, 194
  compulsory, 129–130
  wide gap between cholera epidemics due to, 110
  yellow fever preventive measures, 77, 140
dispensaries, 39
dissemination camps, 123, 145
Dodds, Colonel Amédée, 39
domestic animals
  fees for letting loose in city streets, 122
  ordinance by localists for urban hygiene (1882), 122
  scientific mission findings on urban hygiene, 170
  unsanitary housing shared with, 3–4, 56
drinking water
  ill health linked to polluted, 32
  proposals in cholera aftermath, 114
  sterilization and ozonization of, 171

236 | INDEX

D'Urville, Jacques, 164–165

# E

Echenberg, Myron, 8–9
*Echo de Saint-Louis* (1906–1920), 10
*Ecole du Service de Santé de la Marine*, 10
ecology
　of cholera, 90–93
　of yellow fever, 51–53
El Tor, 1893 cholera epidemic, 110
*"enfants de Ndar"* (sons and daughters of Ndar), 23
*entrepreneurs de travaux publics* (builders), draining swamps, 179–181
environmental factors
　cessation of 1869 cholera epidemic, 102
　local environment theory. *see* localists
　spreading disease in city, 3–4
　wide gap between cholera epidemics, 109–110
epidemic of fear
　influence of medical doctors and, 3
　major disease outbreaks creating, 2
　overview of, 69–72
　pushing for adoption of contagionist strategy, 120
　in yellow fever epidemics, 66
epidemiology of disease
　discovering role of fleas in plague, 186
　identifying health crisis as imperfect knowledge of, 165
　imperfect knowledge of yellow fever, 167
　as main theme of this book, 2
　Martins' account of yellow fever, 134
ethnicity
　distribution of population, 20–21
　Dr. Contaut's proposal for villages based on, 194
　opposition to creation of slums based on, 204–205
　of other groups coming to Saint-Louis, 23–24
　pattern of diseases and, 35

European population
　1904–05 repatriation ordinances, 177–178
　incidence of malaria among, 32
　lacking immunity to yellow fever, 52, 165
　repatriation during 1900 yellow fever epidemic, 65–67
　syphilis affecting, 34
Evaluation Commission, 203–204
evolutionary theory, and documentation, 11

# F

Faidherbe, Governor Léon, 19, 26
Faidherbe Bridge, 74, 154
*faubourgs* (slum neighborhoods). *See also indigènes*
　division of built environment and, 27–28
　fire destroying almost all in 1897, 27
　great sanitation program 1903–1914 and, 181
　photo examples of, 157–158
　spreading infectious agents, 30–31
　urban hygiene in 1882, 122, 124
　urban hygiene in 1901, 170
*faute d'aliment* (for lack of food), and yellow fever, 63, 67–68, 146
fear. *see* epidemic of fear
female *indigènes*
　blamed for carrying disease, 4
　marriage of company agents to local, 18–19
　targeted as source of disease, 72
Ferrand, Député Stanislas, 146–148
Ferrand, Sibaud, 146–147
fevers
　in dry and cold season, 30
　problematic classification of, 31
　from stagnant water, 55
*filles soumises* ("submissive girls")
　medical records on, 9
　scapegoated as carrying disease, 4, 72
Finlay, Dr. Carlos, 52
fire
　destroying thatched-roof huts of *indigènes*, 4
　Guet-Ndar slum easily destroyed by, 170
　purifying Guet-Ndar by, 194–195, 202

fire, *cont.*
   regularly destroying huts of poor in city center, 25
   setting up structures of poor relief from, 19
   threatening built environment, 25–27
Foret, Auguste, 28
Freemasons (*Grand Orient de France*), 28
Freetown, Sierra Leone, 124–127
fumigation. *see* disinfection (fumigation)

## G

garbage collection
   1882 ordinance by localists for, 121
   1901 scientific mission to Senegal and, 171–172
germ theory
   1881 yellow fever epidemic and, 58–59
   microbiology and control of yellow fever, 140–142
   transmission of yellow fever, 56–57
Gokhoumbaye, 24
Grall-Marchoux-Jacquerez Sanitary Mission to Senegal (1901)
   background to mission of inquiry, 164–165
   final report, overview, 166
   follow-up to, 172–173
   investigation, 165–166
   military hygiene, 167–170
   technical issues, 171–172
   urban hygiene, 170–171
*Grande Voirie*, 121
*Grand Orient de France* (Freemasons), 28
Grotte de Lourdes photo, Sor, 161
Guet-Ndar
   1901 scientific mission to Senegal and, 170–171
   decision to rebuild, 204
   development of Saint-Louis at, 24
   fire regulations in, 27
   as long cycle cholera hotbed, 92–93
   stigmatization of urban poor in, 4, 209
Gueye, Aldia, 190–191
Gueye, Fergueye, 196

Gueye family, 23
gum trade
   castes of Wolofs in, 22–23
   decline in, 19, 22
   fortunes made in, 20
   fumigating gum before embarkation, 76
   Moors as difficult partners in, 100–101
Guy, Governor Camille, 176–179

## H

*habitants réfractaires* (rebels), 200–201
Haguenot, 53
health care
   inequality of distribution of water, 94–95
   treatment for malaria and access to, 32
   unequal access to, 31, 35
   yellow fever mortality and access to, 60
Health Service
   creation of autonomic, 43, 78–80
   giving way to medico-military phase of plague policy, 200–204
   not reporting violations to, 179
   radical measures for disinfection, 177–178
   reporting cholera cases to, 110
   sanitary measures for troops and civil servants, 64
HIV/AIDS, new challenge of, 210
*hivernage*-related diseases
   causes of yellow fever, 55
   causing ill health, 31–35
   creation of dissemination camps, 123
   thermal stations for recovery from, 42
*Hôpital Civil*, 37
*Hospice Civil*. *see* Civilian Hospital
hospitals
   dispensaries and other health structures, 41–45
   early period of, 36
   insufficient staffing for epidemics, 43
   medical personnel in, 37–41
   military period of, 36–37
   in public health period, 37
hot climate cliché, 2–3

238 | INDEX

Huot, Dr. L., 181
hygiene, 166, 170–173. *See also* sanitation
Hygiene Colonial Committee, 166
Hygiene Municipal Committee, 166
Hygiene Service (1905), 37

# I

immunity
    of adult survivors of cholera, 92
    developing yellow fever, 165
    implications of assuming racial, 60
    lacking in new host populations in 1878 yellow fever, 61–62
    to malaria, 31–32
    presumption of African, 46
    protective or acquired, 52
    smallpox vaccinations for, 33
incubation period
    cholera, 91
    yellow fever, 51
India, as origin of cholera, 89–90
indictment of leading protestors, anti-plague measures, 196
*indigènes*
    1904–05 sanitation ordinances of Governor Guy, 176–179
    affected by malaria, 32
    believed immunity to yellow fever of, 65, 166
    children lacking immunity to yellow fever, 165
    considered source of disease outbreaks, 56, 70–72
    ecology of yellow fever and, 52
    value judgments about, 11
innoculation (*variolation*), for smallpox immunity, 33. See also vaccination
Institut Pasteur of Dakar (1924), 140
Institut Pasteur of Paris (1888), 140
International Sanitary Conferences, cholera in 1892, 110, 120
intestinal system diseases, from polluted water, 33

isolation
    1900 anti-yellow fever measures, 140
    1902 decree against epidemic diseases, 175
    inadequacies of public health policies for, 43
    state of emergency, August–November 1918, 202–203
    yellow fever preventive measures, 73–76
    yellow fever treatment, 72
isolation camps
    Colonial Hygiene Committee's radical plague measures, 194
    yellow fever mortality in 1981 in, 42
    yellow fever preventive measures, 77–78, 145–146

# J

Jacquerez. *see* Grall-Marchoux-Jacquerez Sanitary Mission to Senegal (1901)
Javouhey, Sister A.-M., 36
*Journal de la Communauté de Saint-Louis (1852–1890)*, 10

# K

Kermorgant, Inspector General of Health Service, 64, 145–146, 148–149
Koch, Robert
    discovery of *Vibrio Cholerae*, 91
    on infectious and parasitical diseases, 140

# L

Lafage, Chief Medical Officer Dr., 131–136
Lailheurgue, Dr., 188
Lampsar project, 105–106
*laptots* (sailors), Wolofs as, 23
lazarettos
    1917 plague outbreak in Saint-Louis and, 187
    contagionsts and quarantine measures, 125

lazarettos, *cont.*
    evacuating people of Guet-Ndar in 1918 to, 202–203
    yellow fever preventive measures, 74
Lazear, Jesse W., 52, 141
*Le Réveil du Sénégal* newspaper, 28
*Les maladies du Sénégal* (Borius), 55
Lévecque, Governor
    biopolitical vs. sanitarian plague strategies and, 190–191
    Blaise Diagne's anti-plague measures and, 193
    radical solution, 197–200
    rejecting radical plague measures, 195–196
    state of emergency, 201–204
Local Committee of Hygiene and Public Health, 177
Local Hygiene Committee, 181
localists
    and contagionists, 58–59
    epidemiological theory of yellow fever and, 55–56
    garbage collection initiatives of, 121
    role of climate in etiology of yellow fever, 59–60
    urban hygiene and, 121–124
    yellow fever preventive measures, 73–74
Loti, Pierre, 30

# M

malaria
    as *hivernage*-related disease, 31–32
    killing more people than yellow fever, 53
    yellow fever vs., 174
*marabouts* (Muslim clerics)
    competition for souls between missionaries and, 113–114
    moderate strategies for plague control, 191
    as most successful Wolofs, 22–23
    protesting 1907 sanitation violation of homes, 179
*marchands* (semi-wholesalers or retailers), 20–21, 23

Marchoux, Dr. Emile. *see* Grall-Marchoux-Jacquerez Sanitary Mission to Senegal (1901); Simond-Marchoux-Selimbeni Mission in Brazil (1901–1905)
maritime quarantine. *see* quarantine, maritime
marriage, of company agents to local women, 18–19
*marseillaises* (vases used as toilets), 39, 171
Martialis, Merlaut, 58
Martin, Raymond, 134–137
M'Boye, Alioune, 196
*Médecin en Chef*
    in 1867 yellow fever epidemic, 61
    Berenger-Féraud's hypothesis on *hivernage*-related disease, 31, 54
    cistern recommendations in cholera aftermath, 114
    of Health Board, 50
    as leader of reorganized Health Service, 79
    Martialis' hypothesis on epidemics, 58–59
    shaping and maintaining colonial order, 80
    threats of epidemics improving ranking of, 43
medical profiling of *indigènes*, 69–72, 121
medic-military phase, plague policy, 200–204
mental illness treatment, 42
*métis*
    believed to have acquired immunity to yellow fever, 80
    emergence of, 19
    as first-class citizens, 23
    parliamentary elections of 1914 and, 29
    rivalries within, 29
    social and political status of, 20–22
miasmatic theory, of yellow fever epidemics
    combining contagionist and localist theories, 58–59
    contagionist school of thought, 56–57
    harmful climate explanations, 54–55
    lack of sanitization, 56
    local environmental explanations, 55–56
    moral emotions leading to, 56
    multi-causal explanations, 55
    physician explanations, 53–54
    racial immunity and acclimatization issues, 60

Military Hospital
  1881 yellow fever epidemic and, 58
  cost of bathing at (1892), 109
  precautions taken in, 73
  protecting health of troops and the elite in, 35
  spread of diseases in, 30–31, 35, 58
  transforming into Colonial Hospital in 1897, 37
  water supply system to, 108, 114
  yellow fever mortality statistics in 1881 at, 62
military period, medical infrastructure (1819–1897), 36–37
missionaries
  assessment of 1893 cholera epidemic, 113–114
  attracting to Saint-Louis, 19
  Bamana as natural allies of, 24
  on panic response to yellow fever, 69
  using archives as information source, 10
  view of epidemics as punishment of God, 51, 101
Moors
  accused of causing cholera, 100–101
  ethnic/racial distribution of, 21
  proposal of ethnic-based village for, 194
  trade relations with, 6, 23
moral emotions, and disease, 56
mortality statistics
  1918 plague, 197–199
  cholera epidemics, 63, 97–100, 112–113
  controlling exposure to *Stegomyia* mosquito to lower, 174
  inadequacy of documentation, 9
  yellow fever epidemics, 51–52, 60, 62–63
mosquitoes
  *Aedes aegypti*, 32, 51, 141–142
  *Aedes Africanus*, 51
  *Aedes Albopictus*, 32
  *hivernage*-related diseases, 31–32
  malarial, 53
  *Stegomyia fasciata*. *see Stegomyia fasciata* mosquito
Municipal Hygiene Service, 177
Muslim
  1893 cholera epidemic and, 113–114

Muslim, *cont.*
  Bamana as enemies of, 24
  clerics. *see marabouts* (Muslim clerics)
  majority of cholera victims as, 100–101
  North part of city as, 28

# N

Navy
  autonomy of Health Service, 78–80
  hospital infrastructure and, 36–37
  medical personnel, 41–44
Ndar Island, 24
Ndar-Toute
  1917 plague outbreak in, 187
  development of Saint-Louis at, 24, 26
  fire regulations in, 27
  long cycle cholera in, 92–93
  migration of infected rats from, 197–198
  mule-drawn cart in (photo), 160
  protesting harsh anti-plague measures, 189
  state of emergency, August–November 1918, 200–204
*négociants* (wholesalers), 20, 22
Network System, role of Saint-Louis in international, 6–8
new paradigm
  1900 yellow fever epidemic and, 142–149
  conclusion, 181–182
  scientific missions to Brazil. *see* Simond-Marchoux-Selimbeni Mission in Brazil (1901–1905)
  scientific missions to Senegal. *see* Grall-Marchoux-Jacquerez Sanitary Mission to Senegal (1901)
noise prohibition, ordinance by localists (1882), 122
nursing care, 36–37, 40–41

# O

overcrowding
  in 1878 yellow fever epidemic, 62

overcrowding, *cont.*
  causing diseases, 3–4, 56, 59
  cholera transmission via, 90, 92–93
  general living conditions of, 30
  in medical institutions, 35–36, 41
  spreading infectious agents, 30–31

# P

Pacini, Filipo, 91
pandemics
  bubonic plague, 186
  cholera, 89–90
panic
  1868 cholera epidemic, 98
  1869 cholera epidemic, 102
  response to yellow fever, 69–72
parliamentary elections of 1914, 29
Pasteur, Louis, 140–141
peanut crops, and plague, 186
pestilential emanations (miasmas), and yellow fever, 56, 58, 74
*Petite Voirie*, 121
pharmacists, early Saint-Louis, 39
physicians
  causes of disease in eighteenth century, 53–54
  Chief Medical Officer. *see Médecin en Chef*
  dispensaries and other health structures, 41–44
  history of first aides to, 36
  influence within colonial administration, 2–3
  Navy, 37–39
  self-perception of colonial, 10
Pinet-Laprade, Governor, 102–103
plague and violence (1917–1920)
  biopolitical strategies vs. sanitarian, 190–191
  Blaise Diagne's intervention, 191–193
  Carrera's intervention, 191
  Colonial Hygiene Committees' radical measures, 193–195
  etiology of bubonic plague, 185–186

plague and violence (1917–1920), *cont.*
  Governor Lévecque's dilemma, 195–196
  mortality, fear, protest, and iradical solution,î 197–200
  outbreak in Saint-Louis, 187–188
  popular protest, 188–190
  state of emergency, August–November 1918, 200–204
  vector control and effective sanitation, 209–210
*Plasmodia falciparum* parasites, 31–32
political history, framework of this book, 5–8
Ponty, Governor General William, 178, 180
"primary" pneumonic plague, 186
Private Council. *see Conseil Privé* (Private Council)
public girls, 72
public health
  inadequacies of policies for, 208–210
  ordinance for urban hygiene (1882), 121–124

# Q

quarantine
  1902 decree on, 175
  inadequacies of public health policies for, 43
  for plague, 187–188, 192–193, 197, 202–203
quarantine, maritime
  for cholera, 89–90, 110
  conflict of interests and, 131–137
  contagionists and, 124–126
  problems from routinization of, 120–121
  protests against, 126–131
  for yellow fever, 61–63, 65, 68, 73–76, 142–143

# R

racial immunity
  1900 yellow fever dissemination camps, 65
  dominating medical thinking, 60
racial susceptibility to colonial pathologies, 143

racism
- 1904 sanitation ordinances and, 177
- 1917 plague outbreak and, 188
- perceptions on seasonal diseases and allergies, 17–18
- radical plague measures and, 194–196
- residential segregation policies based on, 43

rats (*Rattus rattus*), transmission of plague, 185–187, 197
rebels (*habitants réfractaires*), 200–201
Reed, Walter, 141, 173
repatriation of Europeans
- 1900 yellow fever epidemic, 63, 65–67, 145–146
- 1904–05 sanitation ordinances, 177–178
- as French response to turmoil/epidemics, 3
- yellow fever preventive measure, 78

Rio de Janeiro, Brazil
- scientific mission to. *see* Simond-Marchoux-Selimbeni Mission in Brazil (1901–1905)
- unhealthy reputation of, 17

Rocard, Jules, 141
Roume, Governor Ernest, 175–178

## S

sailors (*laptots*), Wolofs as, 23
Sainte Anne dispensary, Ndar-Toute, 41
Sainte-Marie-de-Bathurst, Gambia, 124–126
Saint-Louis, making colonial city of
- built environment, 24–27
- climate, filth and disease, 30–35
- development of trade, 18
- dispensaries and other health structures, 41–44
- founding and spatial layout of, 16–17
- hospitals, 36–37
- tensions and divisions, 27–30
- unhealthy reputation of, 17
- urban society, 18–24

Saint-Philippe, 24
Samb, Makhary, 196
Sanitary Commission, 61, 131–133

sanitary cordons
- 1917 plague, 188
- 1918 plague, 192–193
- Governor General Angoulvant advocating, 197
- yellow fever prevention, 76–77

Sanitary Mission. *see* Grall-Marchoux-Jacquerez Sanitary Mission to Senegal (1901)
sanitation
- 1867 yellow fever epidemic and, 61
- 1882 urban hygiene ordinances, 122
- 1892 cholera epidemic, 111–112
- 1900 yellow fever epidemic and, 64–65, 140–142
- 1901 scientific mission to Senegal and, 166
- 1902 decree for measures, 175
- 1903–1914 great sanitation program, 179–181
- 1904–1907, applying new knowledge, 176–179
- accusations toward *indigènes*, 70–71
- cholera transmission and poor, 92–93
- controlling exposure to *Stegomyia* mosquito, 174
- disease spread from lack of, 3–4
- fifth cholera pandemic, 90
- hygienic measures recommended by CFAO, 143–144
- protests against quarantine measures, 126–131
- race relations and, 4
- radical solution to, 2–3, 197–200
- yellow fever caused from lack of, 56
- yellow fever preventive measures, 73–76

sanitation syndrome
- 1901 scientific mission to Senegal, 169–170, 172
- identifying urban poor quarters with disease, 28
- initiatives since 1882, 121
- stigmatization of *indigènes* and, 72

scapegoating
- conclusions about, 209
- of *indigènes*, 56, 70–72
- long-lasting legacy of explaining disease by, 210
- of submissive girls, 4, 72

scientific missions (1901–1912)
  to Brazil. *see* Simond-Marchoux-Selimbeni Mission in Brazil (1901–1905)
  conclusion, 181–182
  to Senegal. *see* Grall-Marchoux-Jacquerez Sanitary Mission to Senegal (1901)
  seasonal diseases. *see hivernage*-related diseases
Seck, Demba, 179
segregation policies, urban residential
  racialized, 43
  as theme in this book, 2–3
  yellow fever and subsequent, 5
Senegalo-Mauritanian zone, Saint-Louis
  Central Place System model of, 6
  Moors as long-distance traders in, 23
  Network System model of, 7–8
Sengalese nurses, 40–41
separated drainage system, 93
sewerage problems
  1878 yellow fever epidemic and, 61–62
  1901 scientific mission to Senegal and, 167, 171
  1910 great sanitation program for, 179–181
  reducing spread of plague by fixing, 186
  reliance of urban poor on Senegal River water, 43
  as source of ill health, 53
  transmission of cholera via, 90, 92–94
sexually transmitted diseases
  causing ill health in Saint-Louis, 34–35
  female *indigènes* blamed for, 4
  problematic classification of, 31
short cycle (epidemic) cholera, 91–92
*signares* (women entrepreneurs)
  built environment for, 25
  marriage of company agents to, 18–19
  status of *métis*, 19–22
Simond, Paul-Louis. *see* Simond-Marchoux-Selimbeni Mission in Brazil (1901–1905)
Simond-Marchoux-Selimbeni Mission in Brazil (1901–1905)
  applying new knowledge in Saint-Louis, 175–179
  great sanitation program, 179–181
  overview of, 173–174

Sisters (*Soeurs de Saint-Joseph de Cluny*), nursing care of, 36
slaves, creating villages for freed, 24, 26
sleeping sickness, controlling, 41
slum neighborhoods. *see faubourgs* (slum neighborhoods); *indigènes*
smallpox outbreaks, 33–34
Snow, John, 92–93
social system, early, 27–30
Sor
  Bamana and Pulaar settlements in, 23, 26
  Catholic cemetery at, 21, 56, 61, 74
  controlling exposure to *Stegomyia* mosquito in, 178–179
  dispensaries at, 40–41
  Faidherbe Bridge in, 74, 154
  freed slaves living in, 26
  great sanitation program 1903–1914, 179–181
  Grotte de Lourdes (photo), 161
  long cycle cholera in, 92–93
  urban hygiene in 1901 in, 171
  water supply problems in, 95, 107–108
  water tank (photo), 163
sporadic (long cycle) cholera, 91–92
*Stegomyia fasciata* mosquito
  applying new knowledge in 1904–1907, 178–179
  great sanitation program 1903–1914 against, 179–181
  role in yellow fever in 1900, 142
  scientific mission on protection from, 174
  skepticism on findings of, 173
stigmatization of urban poor (*indigènes*)
  1901 scientific mission to Senegal and, 171
  controlling epidemic diseases by, 120, 194–195, 209
  lack of sanitization creating, 4
  moral judgments about, 9–10
  rationalizations for, 25
  rejection of radical plague measures, 195–196
  as theme in this book, 2
  yellow fever and subsequent, 5

submissive girls (*filles soumises*)
    medical records on, 9
    scapegoated as carrying disease, 4, 72
sulfamides, bubonic plague treatment, 186
surgeons, Navy, 37–39
syphilis, 34

## T

Telkamp, G. J., 7
thermal stations, 42
trade
    development in Saint-Louis, 18
    early sanitary cordons interrupting, 76
    ethnicity of local population and, 20
    impact of three major epidemics on, 103–105
    infectious and parasitic diseases spread via, 30
    protests against 1900 anti-yellow fever measures, 137–140
    role of Saint-Louis in Network System of, 6–8
    spread of cholera via. see cholera epidemics
    spread of smallpox via, 33
    spread of yellow fever via, 57
    successful Wolofs and ethnic groups in, 22–23
*Traité des fièvres des pays chauds* (Corre), 124
treatment
    bubonic plague (1896), 186
    cholera, 91, 98–99
    yellow fever, 72–73
tropical anemia cliché, 3
typhoid fever, 33
typhus amaril, *see*

## U

United States Army Commission, 141–142
urban history, framework of this book, 5–8
urban hygiene, 1901 scientific mission to Senegal, 170–171

urban poor. *See also faubourgs* (slum neighborhoods); stigmatization of urban poor (*indigènes*)
    1893 cholera epidemic and, 111–113
    1904 sanitation ordinances and, 177
    biopolitical vs. sanitarian for plague, 190–191
    cholera seen as disease of, 100, 102
    designing new water system for, 105–109
    evacuating people of Guet-Ndar in 1918, 201–202
    forced removals of, 110
    protesting radical anti-plague measures, 188–190
    radical plague measures and, 192–196
    resistance to new knowledge in 1907, 179
    susceptibility to plague, 186
    urban hygiene in 1882, 122
    urban hygiene in 1901, 170
    water supply contamination of, 94–96
urban society, early
    demographics of, 19
    marriage of company agents to local women, 18–19
    other ethnic groups, 23–24
    periods of population growth, 19–20
    problems of, 3–4, 30–31
    social and political status of métis, 19–22
    Wolofs as majority, 22–23
urban system concept, 7

## V

vaccination
    1917 plague outbreak and, 187–188
    1918 anti-plague measures, 193
    1918 state intervention in, 199–200
    Angoulvant advocating, 197
    anti-plague serum, 186
Valière, Governor François-Xavier
    1869 cholera epidemic and, 102
    building sidewalks and drainage in city center, 26
    designing new water system, 105–109
    dissolving Freemasons, 28

Valière, Governor François-Xavier, *cont.*
  distributing water from government cisterns, 94
Vallerstein, Immanuel, 7
*variolation* (innoculation), for smallpox immunity, 33
vases used as toilets (*marseillaises*), 39, 171
ventilation, 30, 72
Veracruz, Mexico, 17
*Vibrio cholerae*, 90–93, 109–110
victim compensation fund, plague, 204
*Village de Segrégation*, Sor, 41
violence, protesting anti-plague measures, 188–190
voting for funds, public works, 144

# W

war
  identifying Moors as enemies, 101
  link between spread of cholera and, 102
waste matter
  1878 yellow fever epidemic, 61–62
  1901 scientific mission recommendations, 170–172
  accusations toward *indigenes* about, 71
  bad odors from, 30
  early efforts to build waterworks, 96
  health challenges of built environment, 43
  ordinance for urban hygiene in 1882, 122
  sewerage problems. *see* sewerage problems
water-borne disease. *see* cholera epidemics
water supply
  1878 yellow fever epidemic, 61–62
  1901 scientific mission recommendations, 171
  1903–1914 great sanitation program, 179–181
  1904 sanitation ordinances, 177
  cholera aftermath and, 114–115
  cholera epidemics and problems with, 90–96
  designing new system for, 105–109
  difficulties of accessing in dry season, 43
  early efforts to build waterworks, 96
  government response to 1868–69 cholera, 101
  ill health caused by polluted, 32–33

water supply, *cont.*
  sewerage problems. *see* sewerage problems
  substantially reducing mortality after 1900, 35
  water tank in Sor (photo), 163
  yellow fever caused by stagnant, 55
"white man's disease." *see* yellow fever epidemics (1867–1900)
Whites, built environment for, 25
winds, yellow fever and "evil," 56
Wolofs
  as majority of early Saint-Louisians, 22–23
  radical plague measures and, 194
  resistance to harsh anti-plague measures, 189
women entrepreneurs (*signares*)
  built environment for, 25
  marriage of company agents to, 18–19
  status of *métis*, 19–22

# Y

yellow fever, developing criteria for identification of, 174
yellow fever epidemics (1867–1900)
  1867, 61
  1878. *see* 1878 yellow fever epidemic
  1880–81. *see* 1880–81 yellow fever epidemic
  in 1899 vs. commercial interests, 131–137
  1900. *see* 1900 yellow fever epidemic
  bacteriology and control of, 140–142
  CFAO recommendations, 142–144
  colonial Health Service, 78–80
  conclusion, 80–81
  contagionists and quarantine measures, 124–126
  ecology of, 51–53
  female *indigènes* blamed as carriers, 4
  as *hivernage*-related disease, 31–32
  lethal character and economic consequences of, 143
  medical personnel and, 37–41
  miasmatic theory and, 53–60
  mortality, 60
  overview of, 50–51

yellow fever epidemics (1867–1900), *cont.*
   panic and medical profiling response to, 69–72
   preventive measures, 73–78
   protests against quarantine measures, 126–131
   toward new paradigm, 145–149
   treatment, 72–73
   as white man's disease, 4–5
Yersin, A., 186
*Yersinia pestis* bacterium, bubonic plague, 185–186